Practising Science Communication in the Information Age

Practising Science Communication in the Information Age

Theorizing professional practices

Edited by

Richard Holliman, Jeff Thomas, Sam Smidt,
Eileen Scanlon, and Elizabeth Whitelegg

Published by Oxford University Press, Great Clarendon Street, Oxford OX2 6DP
in association with The Open University, Walton Hill, Milton Keynes MK7 6AA.

Oxford University Press is a department of the University of Oxford.
It furthers the University's objective of excellence in research, scholarship,
and education by publishing worldwide in

Oxford New York

Auckland Cape Town Dar es Salaam Hong Kong Karachi
Kuala Lumpur Madrid Melbourne Mexico City Nairobi
New Delhi Shanghai Taipei Toronto

With offices in

Argentina Austria Brazil Chile Czech Republic France Greece
Guatemala Hungary Italy Japan Poland Portugal Singapore
South Korea Switzerland Thailand Turkey Ukraine Vietnam

Oxford is a registered trade mark of Oxford University Press
in the UK and in certain other countries

Published in the United States
by Oxford University Press Inc., New York

First published 2009

© The Open University 2009

British Library Cataloguing in Publication Data
Data available

Library of Congress Cataloging in Publication Data
Data available

Typeset by Graphicraft Limited, Hong Kong
Printed in Great Britain
on acid-free paper by
Ashford Colour Press Ltd, Gosport, Hampshire

This book forms part of the Open University course SH804 *Communicating science in the
information age*. Details of this and other Open University courses can be obtained from
the Student Registration and Enquiry Service, The Open University, PO Box 197, Milton Keynes
MK7 6BJ, United Kingdom: tel. +44 (0)845 300 60 90, email general-enquiries@open.ac.uk

http://www.open.ac.uk

ISBN: 978-0-19-955267-2

■ PREFACE

This volume—similarly its companion (Holliman *et al*. 2009)—deals with developments that have occurred in science communication over the last 10 years, in this case in relation to the study of professional practices. During this time we have continued to teach a previous Open University course called *Communicating Science* (see Scanlon *et al*. 1999a,b), as part of the MSc in Science, MSc in Science and Society and Postgraduate Diploma in Science and Society.

The study and practices of science communication have developed quite dramatically during this time. We are therefore eager to acknowledge that this volume is, at least in part, informed by these developments, by our experiences of producing (and presenting for 10 years) *Communicating Science*, and by our interactions with a wide range of students, many of whom have been practising science communicators, some of whom have gone on to become researchers and practitioners in this burgeoning field of scholarship.

This volume has a particular focus on the practices of contemporary science communication. Our planning for this volume has been informed by what we have learnt as practitioners of science communication, principally as academics teaching and researching these issues. For example, members of the editorial team (RH and JT) have worked as part of the European Network of Science Communication Teachers (The ENSCOT Team 2003), and made contributions to the literature in terms of researching how geographically distributed students have engaged collaboratively to learn about media reporting of science (Holliman and Scanlon 2006). More practically, we have also acted as peer reviewers of science communication research, and grant and book proposals, and made contributions to a range of media forms, including collaborations with BBC Radio 4's *Material World* magazine programme and to http://www.open2.net and the ISOTOPE web portal (http://isotope.open.ac.uk).

We are grateful to a number of academics who have supported our efforts in the past, including: our two respective external examiners, Roger Hartley and Brian Trench; external assessors on the first course and this one (Robin Millar and Joan Leach, respectively); guest lecturers, including Jane Gregory, Alan Irwin, Joan Leach and Brian Trench; members of the original production course team, including Roger Hill, Jill Tibble (BBC), Carol Johnstone, Kirk Junker, Hilary MacQueen, Cheryl Newport, Shelagh Ross, Rissa de la Paz (BBC) and Simeon Yates; and the support from Peter Taylor and Susan Tresman who each led the postgraduate programme offered by the Science Faculty at the Open University over this period.

No course of this nature would be possible without a committed and skilled support team. To this end, we are especially grateful to the ongoing efforts of Christine Marshall and Carol Johnstone, and to James Davies and Martin Chiverton. Furthermore, we are also grateful to Jonathan Crowe from Oxford University Press and Christianne Bailey from the OU's co-publishing department for their support in the development of this project, also to Giskin Day, Clari Hunt, Shelagh Ross and Jane Perrone.

We acknowledge the following for permission to use photographs for the Section and Chapter openings. In Chapter 2.1 the photograph of the symposium coffee break was provided by Christine Marshall and features Jeff Thomas and Pat Murphy in the foreground. (For more information about this symposium see http://www.open.ac.uk/science/SEH806/Symposium/.) In Chapter 5.1 the photograph of the book group was provided by Elizabeth Whitelegg. In Chapter 5.3 the photograph of Jon Dennis at work in *The Guardian* podcast studio was provided by Jane Perrone. (For more information about *The Guardian* science podcasts see http://www.guardian.co.uk/science/series/science.) In Chapter 6.1 the photograph of *Our Dynamic Earth* was provided by Stuart Monro. (For more information about *Our Dynamic Earth* see http://www.dynamicearth.co.uk/.) In Chapter 6.2 the photograph of the Café Scientifique was provided by Ann Grand. (For more information about Café Scientifique see http://cafescientifique.org/.)

Finally, we would also like to acknowledge those who have worked at the 'coal-face' of Open University teaching, the course tutors; the course would not have been a success without them. Over the years we have worked with a number of academics who have taken on this role. In alphabetical order they are: Kim Alderson, Susan Barker, Bruce Etherington, John Forrester, Richard Holliman, Vic Pearson, Charlotte Schulze, Zbig Sobiersierski, Rachel Souhami, Anna Tilley and Simeon Yates.

Richard Holliman,
Jeff Thomas,
Sam Smidt,
Eileen Scanlon and
Elizabeth Whitelegg

■ REFERENCES

Holliman, R. and Scanlon, E. (2006). Investigating co-operation and collaboration in near synchronous computer mediated conferences. *Computers & Education*, 46(3), 322–35.

Holliman, R., Whitelegg, E., Scanlon, E., Smidt, S. and Thomas, J. (eds) (2009). *Investigating Science Communication in the Information Age: Implications for Public Engagement and Popular Media*. Oxford University Press, Oxford.

Scanlon, E. Hill, R. and Junker, K. (eds) (1999a). *Communicating Science: Professional Contexts*. Routledge, London.

Scanlon, E., Whitelegg, E. and Yates, S. (eds) (1999b). *Communicating Science: Contexts and Channels*. Routledge. London.

The ENSCOT Team (2003). ENSCOT: the European Network of Science Communication Teachers. *Public Understanding of Science*, 12(2), 167–81.

■ CONTENTS

ABBREVIATIONS AND ACRONYMS   ix

BIOGRAPHIES OF CONTRIBUTORS   x

INTRODUCTION TO THE VOLUME   xv

SECTION 1 **Communicating post-academic science**   1

1.1   Scientists communicating, *by Jane Gregory*   3

1.2   Ethical codes and scientific norms: the role of communication in maintaining the social contract for science, *by Robert Doubleday*   19

1.3   Patents and the dissemination of scientific knowledge, *by Charlotte Schulze*   35

SECTION 2 **Developing trends in scientists' communicating**   51

2.1   Science communication across disciplines, *by Joachim Schummer*   53

2.2   Communicating physics in the information age, *by Matthew Chalmers*   67

SECTION 3 **Accessing contemporary science**   81

3.1   Science and the online world: realities and issues for discussion, *by Scott L. Montgomery*   83

3.2   From print to online: developments in access to scientific information, *by Richard Gartner*   98

SECTION 4 **Consensus and controversy**   113

4.1   Peer review in science journals: past, present and future, *by Elizabeth Wager*   115

4.2   Controversy and consensus, *by Jeff Thomas*   131

SECTION 5 **Popularizing science**   149

5.1   Where do books fit in the information age?, *by Bruce V. Lewenstein*   151

5.2   Science communication in fiction, *by Jon Turney*   166

5.3   Speaking to the world: radio and other audio, *by Martin Redfern*   178

SECTION 6 **Practising public engagement** 193

6.1 The development of *Our Dynamic Earth*, *by Stuart Monro* 195

6.2 Engaging through dialogue: international experiences of Café Scientifique, *by Ann Grand* 209

FINAL REFLECTIONS . . . 227
INDEX 231

ABBREVIATIONS AND ACRONYMS

AAAS	American Association for the Advancement of Science	**ISDN**	Integrated Services Digital Network
ALPSP	Association of Learned and Professional Society Publishers	**IVF**	*In vitro* fertilization
		JAMA	*Journal of the American Medical Association*
ABSW	Association of British Science Writers	**JANET**	Joint Academic NETwork
BA	British Association (for the Advancement of Science)	**JISC**	Joint Information and Systems Committee
		LHC	Large Hadron Collider
BBC	British Broadcasting Corporation	**NGO**	Non-governmental organization
BGS	British Geological Survey	**NIH**	National Institutes of Health
BSE	Bovine spongiform encephalopathy	**NNI**	National Nanotechnology Initiative
BMJ	*British Medical Journal*	**NSF**	National Science Foundation
CERN	Organisation Européenne pour la Recherche Nucléaire	**NSTC**	National Science and Technology Council
		OECD	Organisation for Economic Co-operation and Development
COPE	Committee on Publication Ethics	**PCR**	Polymerase chain reaction
CORE	Comment on Reproductive Ethics	**RAE**	Research Assessment Exercise
DOAJ	Directory of Open Access Journals	**RAJAR**	Radio Joint Audience Research
DTI	Department of Trade and Industry	**ROAR**	Registry of Open-Access Repositories
EASE	European Association of Science Editors	**SLAC**	Stanford Linear Accelerator Center
EPO	European Patent Office	**SMC**	Science Media Centre
EU	European Union	**TRIPS**	Trade-related aspects of intellectual property rights
FTP	File transfer protocol		
GM	Genetically modified	**UN**	United Nations
HFEA	Human Fertilisation and Embryology Authority	**URL**	Uniform resource locator
HGA	Human Genetics Alert	**USPTO**	United States Patent and Trademark Office
IFLA	International Federation of Library Associations	**WAME**	World Association of Medical Editors
IP	Intellectual property	**WHO**	World Health Organization
IPCC	Intergovernmental Panel on Climate Change	**WIPO**	World Intellectual Property Organization
IPR	Intellectual property rights	**WTO**	World Trade Organization

■ BIOGRAPHIES OF CONTRIBUTORS

Matthew Chalmers completed a physics degree at Glasgow University in 1996, then spent 3 years at the European laboratory CERN near Geneva where he completed a PhD in experimental particle physics. In parallel with this research (measuring the mass of the 'W boson') he became increasingly involved in outreach activities and, after a short post-doc, decided to pursue science communication as a career. He won an ABSW/Wellcome Trust bursary which allowed him to undertake an MSc in science communication at Imperial College, starting in 2001. After a work placement at *New Scientist* he got the job of features editor at *Physics World* in Bristol in November 2002. He left the magazine in December 2007 to pursue a freelance career in science journalism and media training.

Robert Doubleday is a Research Associate in the Department of Geography at the University of Cambridge. His research is concerned with the relationship between science, citizenship and the governance of emerging technologies. He is currently working on a project to develop methods for articulating the public dimensions of research on nano-technologies as applied to the biomedical sciences. Prior to joining the Department of Geography, he was based in the University of Cambridge's Nanoscience Centre, where he worked collaboratively with scientists to explore the societal aspects of nanoscience.

Richard Gartner is an information professional who has specialized in the fields of electronic information provision for over 20 years. From 1991–2007 he was New Media Librarian for Oxford University Libraries, where he was responsible for the introduction of the internet into the Bodleian Library, the library's first CD-ROM network and its first digital imaging projects. In recent years, he has specialized in metadata for digital libraries, in which capacity he is a member of the editorial board for the Metadata Encoding and Transmission Standard (METS) for digital library metadata. He has a professional interest in the role of digital libraries in international development, and has visited several libraries in the developing world in his professional capacity. He is also an Open University student, currently working towards an MA in music.

Ann Grand is Assistant Organizer for Junior Café Scientifique. As part of this role, she looks after the world-wide Café Scientifique family; acting as a central hub for information, supporting new café organizers and managing the web site. In 2003, she founded the Bristol Café Scientifique. Her other jobs include being information systems manager for a local school, a freelance book editor and Company Secretary of Cyberlife Research Ltd., founded by Ann and her ex-husband to carry out research into artificial life. Ann trained as a biology teacher and later took an Open University degree, studying science topics from geology, to systems theory, to the politics of health care.

Jane Gregory is Senior Lecturer in Science and Technology Studies at University College London. Her research interests include the public dimensions of science, unorthodox science and the history of recent science. She is co-author of *Communicating Science* (Longman, 1991) and *Science in Public* (Plenum, 1998) and author of *Fred Hoyle's Universe* (Oxford University Press, 2005).

Richard Holliman is Senior Lecturer in Science Communication at the Open University (OU), UK and production course team chair of *Communicating Science in the Information Age*. After completing a PhD investigating the representation of contemporary scientific research in television and newspapers in the Department of Sociology at the OU, he moved across the campus to the Faculty of Science. Since that time he has worked on a number of undergraduate and postgraduate course teams, and as part of the European Network of Science Communication Teachers (ENSCOT), producing mixed-media materials that address the interface between science and society. He edited (with Eileen Scanlon) *Mediating Science Learning Through ICT* (2004, Routledge) and, more recently (with Jeff Thomas) a special issue of the *Curriculum Journal* (**17**: 3) on science learning and citizenship. He is a member of the OU's Centre for Research in Education and Educational Technology and is leading (with colleagues) the ISOTOPE (Informing Science Outreach and Public Engagement) and *(In)visible Witnesses* research project teams.

Bruce V. Lewenstein is Professor of Science Communication in the Departments of Communication and of Science and Technology Studies at Cornell University, Ithaca, New York, USA. He works primarily on the history of public communication of science, with excursions into other areas of science communication (such as media coverage of emerging technologies and sociological aspects of open-access publishing). He has been active in international activities that contribute to education and research on public communication of science and technology, especially in the developing world. He is a co-author of *The Establishment of American Science: 150 Years of the AAAS* (Rutgers University Press, 1999, with Sally Gregory Kohlstedt and Michael M. Sokal), editor of *When Science Meets the Public* (AAAS, Washington, DC, 1992), and co-editor of *Creating Connections: Museums and the Public Understanding of Research* (Altamira Press, 2004, with David Chittenden and Graham Farmelo). From 1998 to 2003, he was editor of the journal *Public Understanding of Science*. He is a Fellow of the American Association for the Advancement of Science.

Stuart Monro, Scientific Director of *Our Dynamic Earth*, is a geologist who graduated with a first-class honours degree at Aberdeen University in 1970. He has a PhD from the University of Edinburgh. He spent most of his career with the British Geological Survey (BGS) working on sedimentary rocks, the evolution of sedimentary basins and on environmental geology. He retains academic links with the Open University where he is an Associate Lecturer in Earth Sciences and with Edinburgh University where he is Visiting Professor in the School of Geoscience. Within *Our Dynamic Earth*, Stuart has been responsible for the scientific content of the facility, formulating the interpretative plan for the exhibition and working with the exhibition designers to ensure the scientific accuracy of the exhibition. He is a trustee of the National Museums Scotland, a non-executive director of the Edinburgh International Science Festival and independent co-chair of the Scottish Science Advisory Committee.

Scott L. Montgomery is a petroleum consultant, author and adjunct faculty member in the Jackson School of International Studies, University of Washington, USA. He has written widely on frontier exploration issues, particularly in North America, and is currently completing a book on energy issues in the 21st century. His most recent book is *The Chicago Guide to Communicating Science* (University of Chicago, 2003). He holds a BA in English from Knox College and an MS in geological sciences from Cornell University.

Martin Redfern joined the BBC as a studio manager in 1975, following a geology degree at University College London. He joined the World Service Science Unit in 1981, working initially as a writer and producer. Apart from a brief venture into television in 1986, when he worked on the BBC magazine programme *Tomorrow's World*, he has remained in radio science ever since, being variously described as senior producer, assistant editor, executive producer and chief producer. In 1999 the unit merged with the Radio 4 Science Unit. The BBC Radio Science Unit now makes programmes for both the World Service and Radio 4.

Eileen Scanlon is Professor of Educational Technology at the Open University and a Visiting Professor at the Moray House School of Education, University of Edinburgh. She co-directs the Centre for Research in Education and Educational Technology at the Open University. Her books include *Communicating Science: Professional Contexts* (Routledge, 1999), edited with Roger Hill and Kirk Junker; *Communicating Science: Contexts and Channels* (Routledge, 1999) edited with Elizabeth Whitelegg and Simeon Yates; *Reconsidering Science Learning* (Routledge Falmer, 2004) edited with Patricia Murphy, Jeff Thomas and Elizabeth Whitelegg; and *Mediating Science Learning with ICT* (Routledge Falmer, 2004) edited with Richard Holliman.

Charlotte Schulze was educated at the universities of Munich, Konstanz, Freiburg and London. In addition to an MSc in history of science and a PhD in cell biology from Imperial College, she holds an MSc in management of intellectual property from Queen Mary and Westfield College, all in London. She is an Associate Lecturer and Associate Teaching Fellow at the Open University.

Joachim Schummer is Heisenberg Fellow at the University of Darmstadt. After double-graduation in chemistry and philosophy, and a PhD and Habilitation in philosophy at the University of Karlsruhe, he has held teaching and research positions at the University of South Carolina, University of Darmstadt, Australian National University and University of Sofia. His research interests focus on the history, philosophy, sociology, ethics and public understanding of science and technology, with an emphasis on chemistry and, since 2002, nanotechnology. His recent book publications include *Discovering the Nanoscale* (2004, 2005), *Nanotechnology Challenges* (2006), *Nanotechnologien im Kontext* (2006) and *The Public Image of Chemistry* (2007). He is the founding editor of *Hyle: International Journal for Philosophy of Chemistry* (since 1995) and serves on various international committees, including the UNESCO expert group on nanotechnology and ethics.

Sam Smidt is a senior lecturer based in the Department of Physics and Astronomy at the Open University, and award director of the MSc in Science. She has contributed to a number of courses in physics and also to courses concerned with contemporary science and society issues at both undergraduate and masters level, including *Science in Context*, *Contemporary Issues in Science Learning* and *The Science Project Course: Science and Society Project*. She has interests in physics education and outreach work that promotes science to the public.

Jeff Thomas is a senior lecturer within the Department of Life Sciences at the Open University (OU). He has worked at the OU all his professional life, contributing to a wide range of teaching initiatives in biology and in health sciences, and more recently to a range of projects concerned with contemporary science issues and on the relationships between science and different publics, at both undergraduate and Masters level, and as part of the European Network of Science Communication Teachers (ENSCOT). He co-edited *Science Today; Problem or Crisis* (with Ralph Levinson; Routledge, 1997) and *The Sciences Good Study Guide* (with Andrew Northedge, Andrew Lane and Alice Peasgood; Open University, 1997), and (with Richard Holliman) co-edited a special issue of the *Curriculum Journal* (**17**: 3). His research interests are concerned with the influence of contemporary science controversies on public attitudes, on conceptual problems of learning biological science and in public involvement in science-based policy-making. He also teaches part-time for Birkbeck College, University of London on its Diploma in Science Communication.

Jon Turney is a science writer, editor and lecturer based in Bristol, UK, who has published widely on popular science and science communication. From 2005–08 he led the MSc in creative non-fiction at Imperial College London. His books include *Frankenstein's Footsteps: Science, Genetics and Popular Culture* (Yale, 1998). He wrote the chapter on popular science books for the *Handbook of Public Communication of Science and Technology* (Routledge, 2008). His current project is *The Rough Guide to the Future*, due out in Autumn 2009.

Elizabeth Wager studied zoology at Oxford and Reading Universities and then worked in the publishing and pharmaceutical industries ending up as head of international medical publications for Glaxo-Wellcome. Since 2001 she has run her own company, Sideview, which provides training, writing and publication consultancy services. She has run courses for doctors, journal editors and medical writers on five continents. She is the author of books on *Getting Research Published: an A to Z of Publication Strategy* (Radcliffe, 2002) and *How to Survive Peer Review* (BMJ Books, 2002; with Fiona Godlee and Tom Jefferson). She is a co-author of *Good Publication Practice for Pharmaceutical Companies*, the European Medical Writers Association guidelines on the role of medical writers, the *Blackwell Best Practice Guidelines on Publication Ethics*, and Cochrane systematic reviews on the effects of peer review and technical editing. She is Secretary of the Committee on Publication Ethics (COPE), and a member of the *BMJ* Ethics Committee and the editorial boards of *European Science Editing* and the *Medscape Journal of Medicine*.

Elizabeth Whitelegg is Senior Lecturer in Science Education in the Science Faculty at the Open University, and inaugural Award Director for the Science Short Course Programme which she continues to lead. She has worked on a range of science and society-related courses, including *Communicating Science* and *Contemporary Issues in Science Learning*. As a member of the Centre for Research in Education and Educational Technology her main research interest is in girls' and women's participation in science and in learning science (particularly physics) at all levels and she recently produced (with Professor Patricia Murphy) a review of the research literature on the participation of girls in physics commissioned by the Institute of Physics. She is currently leading (with Richard Holliman) the *(In)visible Witnesses* project. In 2003 she was invited to become a Fellow of the Institute of Physics.

■ INTRODUCTION TO THE VOLUME

In this introduction to the volume we offer some advice on how we feel the chapters could be used by students, practitioners, researchers and teachers of science communication, and provide an overview and rationale for the selection of the overall theme of the book, its sections and the chapters therein.

How to use this book

This book explores contemporary practices of science communication; the plural is deliberate. There is no one way to practice science communication, no simple 'one-size-fits-all' solution to these complex, sometimes fascinating, but often challenging issues. To become better at communicating science we argue that there is a strong rationale for practitioners to reflect critically on existing and emerging practices. Even then, successful science communication is rarely conclusively demonstrated.

When we began to commission chapters for this edited collection we started from the premise that a wide range of practitioners with differing forms of 'professional', 'pro-am' (Leadbetter and Miller 2004) and 'citizen' expertise practice the communication of contemporary science in various forms and contexts, influenced by a range of different motivations and attendant constraints. We wanted to reflect this diversity in the selection of contributing authors. Others maintain a scholarly interest in studying these practices, and we wanted to complement practitioner perspectives with a selection of research-informed accounts. Researchers, of course, have important things to say about practices of science communication, but informed by evidence, rather than experience. It follows that what binds these authors together, if only in terms of their contributions to this book, are their interests in the practices of science communication and a willingness to engage in critical reflection about them. In this sense a key aim of the book is to contribute to the ongoing theorizing of these practices.

As with the companion volume to this book (Holliman *et al.* 2009) we see great strengths in inviting students (practitioners, researchers and lecturers) of science communication to explore diversity in science communication, in this case in terms of its practices, and the study thereof; to begin to compare and contrast how practitioners and researchers have considered the communication of science, and to note how and why practitioners differ in their approaches. As with so many things in life this book illustrates that the enactment of particular practices 'could be otherwise'. We argue that a greater understanding of the question of 'why is science communication as it is?' can inform practices in valuable ways. In this sense, we wish to engender critical engagement with what we (and many others) consider to be important and often complex issues.

Of course, we do not claim to have addressed everything that a reader might want to know about the practices of contemporary science communication. Instead, our aim has been to commission and edit what we consider to be stimulating and thought-provoking chapters from leading practitioners and researchers, and to encourage the use of this book as a starting point to then explore the landscape of science communication practices in more detail.

The structure of the book

The book is split into six sections, each considering a particular theme relevant to the contemporary practices of science communication.

Section 1: Communicating post-academic science

What unites the three chapters that comprise this section is their focus on the practices of communicating 'post-academic' science. It is argued that a sustained phase of cultural evolution has now positioned practitioners of science communication firmly within the 'post-academic', or 'mode 2', landscape of science (Gibbons *et al.* 1994). Science nowadays is more influenced than hitherto by commercial, political and social forces—or at least such influences are more evident and more worried about. How science is communicated has played a critical role in bringing about such changes and also in consolidating and extending its influence.

Jane Gregory writes perceptively of modern-day communication practices in the first chapter in this section, looking at how communications technology has affected 'traditional' working practices of science and scientists. She shows (as do other authors in this volume) that amidst the welcome opportunities for change, long-established practices—notably those of formal publication and peer review—doggedly linger on, perhaps a reflection of their proven worth. But if scientists in their parallel lives—their social and family settings—are absorbing additional communication skills, these new tools may also find their way into professional communication practices. And as they do so, the character of the scientific community is changed, raising new questions of the type Gregory explores, about how notions of distance and 'separateness' of the community from non-scientists might alter as a result.

Robert Doubleday's chapter focuses on Robert Merton's 'CUDOS' norms and the extent to which they are relevant to the practices of contemporary science communication. Merton's norms arguably offer a description of the social characteristics of academic science as they were in the 1940s and in today's contemporary post-academic climate they have an appealing innocence and poignancy for a scientific age that was always more imagined than real. But Doubleday goes beyond critiquing Merton's norms as first formulated, and looks at their appropriateness in today's post-academic climate. His arguments have a strong bearing on what is sometimes termed the 'deficit of trust' suffered by contemporary science (see Irwin 2009). What sort of knowledge and scientific ethos inspires and deserves trust? Ensuring a greater openness for science, with porous boundaries encouraging the free flow of communication, would be a precondition. Using a tangible and contemporary example, Doubleday looks at the prospect of establishing an ethical code of a type both workable and ambitious in scope—one that is consistent with the inherent competitiveness that helps drive post-academic science forward.

A form of 'industrialization' of science is a further diagnostic feature of post-academic science, and in the third chapter in this section the focus is on the province of patenting law and maximizing commercial opportunity. Charlotte Schulze's chapter takes an imaginative look at the key notion of 'intellectual property' and at patents in particular. This is achieved via a focus on a case study of a genetically modified plant, drawing comparisons between the types of communication evident in the formal peer-reviewed publication and the associated patent application. For the post-academic scientist, the requirement to craft patent applications and navigate bureaucratic obstacles as demanding as the worst excesses of laboratory politics places a further set of communication imperatives that make life both more demanding and rewarding than hitherto. What Schulze, Doubleday and Gregory illustrate through their respective chapters is that post-academic science clearly requires communication skills of a distinctive range and quality.

Section 2: Developing trends in scientists' communicating

Two chapters are included in this section. The first, by Joachim Schummer, examines the issues faced by scientists communicating with scientists of other disciplines in interdisciplinary science projects. Schummer considers the way in which science has fragmented into an ever-larger number of disciplines and discusses what defines a discipline in terms of both cognitive and social aspects. He then explores strategies that facilitate communication across disciplines, again in both cognitive and social terms. The chapter concludes with a look at nanotechnology as a contemporary example of an interdisciplinary area, and the communication issues raised by complex interdisciplinarity and by the politics driving forward the nanotechnology agenda.

The second chapter in this section, by Matthew Chalmers, looks at the way contemporary communication issues are being managed within the physics community. With much of the early internet and web development taking place within physics, physicists were amongst the most enthusiastic early adopters of these forms of electronic science communication. Chalmers looks back at how physicists have used the web to communicate primary scientific literature, for example through examination of services such as the arXiv preprint server, thereby pushing the boundaries of information sharing over the last 20 years. He also examines the developing web 2.0 applications and finds physicists to be more reluctant adopters of user-generated content and online collaborative authoring than many non-scientists. Nonetheless there is a growing trend for physics blogs, and these are being used for communication within the discipline as well as to communicate with the wider world. The paper also looks at wikis and at some of the online social networking and video-sharing sites, briefly discussing how these are starting to be used within the physics community. But Chalmers' chapter does more than describe new communication channels—it suggests that these new opportunities will inevitably change how physics (and science) is undertaken and communicated—not just the types of questions asked but what type of knowledge is generated and how it is argued over.

Section 3: Accessing contemporary science

Section 3 comprises two chapters. Both examine how well-established and emerging ways of accessing scientific literature is shaping how contemporary science is conducted.

Scott Montgomery's chapter gives a perspective on the communication of science stretching from the distant past to the near future. He emphasizes the way in which the forms of communication available to scientists change the way in which science is communicated and that each change can be traced to the social and technological influences that have shaped them. He draws particular attention to the way that the internet has changed how science is done, calling it 'a many-splendoured tool in the politics of knowledge'. He notes the way that the interaction between researchers has changed the relationship between authors and readers. One surprising feature is the opportunity given to readers to comment directly on published materials. The increased interaction between scientists and their publics is very marked in the increased involvement with outreach and engagement activities undertaken by research centres, government organizations and companies which have been enormously expanded by the internet. Interaction between scientists has also changed: transfer of data between collaborators, the expansion of the visual dimension by the use of graphics, simulations and modelling, and access to electronic journals. He also draws attention to some potential disadvantages in terms of 'leaked data' and increased (undesirable) publicity. Rather than assuming a simplistic view, that the internet is necessarily a force for good, he concludes that the influence of a new technology needs to be considered in terms of existing social relations and dependencies. Therefore, although the internet is a 'boon to science', its capabilities will take time to be fully explored.

Richard Gartner's chapter takes the specific issue of the impact of electronic provision of information on how information is accessed. The chapter begins by looking at the main medium which researchers have traditionally used; how print was used to access the scholarly record in the form of monographs, journals and transactions of learned societies. Access to such information was influenced by the production of indexes and abstracts. The first change in ease of access came with the move to digital data bases, but these were not networked. A more significant shift was the most recent moves which have been made to allow access to the academic record online. Gartner draws attention to the fact that the model of academic publishing remains remarkably stable despite these shifts, noting the continuing primacy of the peer-reviewed article in an academic journal as the 'gold standard' of verified knowledge, at least in terms of the judgement of many bench scientists. He also discusses moves towards greater openness by charting the move to open-access journals and institutional repositories. He questions whether author engagement with these new initiatives is influenced by partial perceptions of the readership and somewhat artificial citation levels of such journals and the papers they contain.

Section 4: Consensus and controversy

This section comprises two chapters that neatly juxtapose two contrasting features that drive science forward in different ways: consensus and controversy. The first chapter is about a key feature of how standards are judged in science—how, by consensual agreement, particular findings (book proposals or grant applications) are deemed fit for purpose. The second is about contemporary examples of scientific conflict—of evidence, of theory, of opinion, of ethical stance or some often explosive mix of all such factors—and how scientists can deliberately manage such situations, often under full public gaze.

Elizabeth Wager's chapter offers a balanced and insightful commentary on the practices of peer review. Her major focus is on this longstanding art in relation to the publication of research findings in scholarly medical journals. She addresses 'big' questions such as: how is peer review now undertaken, is it a fair process, and what are its prospects? The issue of how peer review will work in the 'information age' is central to this book's overall theme. The long-standing mechanism for the dissemination of scientific knowledge—the printed word on paper—may seem antediluvian a few years hence. Were it not for the massive publishing interests mentioned by Jane Gregory in this volume, and the admirable predilection of scholars and students for the delights of black ink on paper, professional life would be even more screen-based than at present. Given the present-day ease of dissemination, should the process of authentication of scientific findings be opened up, to be more transparent, more egalitarian, quicker, fairer? Wager's conclusions about the future prospects for peer review are offered with the benefit of her wide experience; the prospects for allying new technologies with well-established practices of authentication may be better than many might first suppose.

Controversies in science are considered in Jeff Thomas' chapter. Those of particular interest come to public attention through acts, statements and proposals from scientists that selectively draw interest from news media. Rather than a comprehensive review, his approach is to look at a few examples representative of more general contemporary trends. Through such examples he argues that if there ever were a time when science was dominated by practitioners shy of the public spot-light then those days are long gone—and unmissed. Nor are scientists hapless victims of manipulative mass media, frozen in the unkind glare of publicity. Scientists are now more able and adept at creating self-interested agendas, at lobbying and cultivating a mix of ordinariness and distinction designed to appeal and affect. Modern communication technologies can more readily mobilize the 'scientific voice' to concerted action. Years of encouragement to scientists to understand more of the workings of 'the media', often by means of specially designed training courses in 'how to handle the media', are resulting in greater numbers of 'media savvy' scientists that show the appropriate degree of acumen and facility for newsworthy 'sound bites'. The communication skills on show are very different from those practised in what earlier generations have assured us were more genteel times.

Section 5: Popularizing science

The three chapters in this section examine the popularization of science within very different contexts—that of the popular science book, the novel and of 'multi-platform audio' (previously known as wireless—which itself has taken on another meaning for 'mobile' practitioners and researchers of science communication—and now more commonly referred to as radio, podcasts, downloads and so on). All three chapters consider who the audiences for these various forms of science communication are, and the qualities needed within these forms to encourage successful communication.

In the first chapter in this section, Bruce Lewenstein provides an analysis for the continued place of science books as a vehicle for popularizing science. He provides a strong defence for the continued place of printed books by demonstrating the complementarity of science books in printed form and electronically delivered resources that extend the value of books to the readers. The chapter considers a wide range within the genre of non-fiction science books, from textbooks, initially for science learners and research scientists (but now also reaching a wider 'pro-am' readership), to those classed as popular science books that are the subject of cultural discussions amongst scientists and non-scientists. In discussing the purposes of these, the chapter examines the state of health of C. P. Snow's perennial 'two cultures' divide between the arts/humanities and the sciences and how the publication of 'breakthrough books' (such as Bronowski's *Ascent of Man* and Sagan's *Cosmos* and the attendant television series), played a central role in engendering public discussion about these publishing phenomena. The continuation of this trend with the publication of such books as Dawkins' *The God Delusion* remind us yet again of the important and continuing role of the printed medium within the field of science communication.

In his chapter about science-in-fiction novels (as opposed to science fiction novels), Jon Turney considers whether this form of fiction can communicate science. He critiques several examples to illustrate the main characteristics that need to be present in order for some form of science communication to occur. He concludes by suggesting that a viable goal of a work of science-in-fiction, if it is to be considered as a vehicle for communicating science, would be when it allows for exploration of the future possibilities and implications of established scientific ideas. This can occur at a time when these possibilities and implications can only be guessed at with the level of current scientific knowledge available.

In Martin Redfern's chapter he describes his personal journey through a long career in science radio broadcasting. He charts the development of radio from single-platform broadcasting over the airwaves (wireless and analogue radio) to the situation today with multi-platform audio, using various digital technologies. He describes the shifts over time in the production processes and values according to the developing needs, locations and demands of producers and audiences, noting how changes—for example, with the development of new technologies and with raised expectations of audiences—have influenced the style and nature of established and emerging forms of audio communication.

Section 6: Practising public engagement

The two chapters in the final section of the book present practitioner narratives about the evolution and development of two different sites for 'public engagement', respectively, using this term in a broad sense.

In the first chapter, Stuart Monro, the scientific director of *Our Dynamic Earth*—a hands-on science centre built as a millennium project—tells the story of the choice of the particular location in Edinburgh and the interaction of this location with the centre's environmental theme. He describes how the historic legacy of the city's industrial heritage, combined with the surrounding geology of the site, has resulted in a science centre that aims to bring an understanding of earth and environmental sciences to an audience ranging from young children to the elderly in an entertaining way, whilst still maintaining its primary educational purpose.

A stimulating perspective is given by the second chapter which charts the growth of the Café Scientifique movement and its development within the UK and internationally. Ann Grand argues that Café Scientifique seeks to maintain links between science and culture, offering informal venues where expert scientists can meet *with* publics to discuss and debate contemporary scientific issues outside traditional arenas for these events, such as university lecture halls. The various models for Café Scientifique events world-wide are presented, including the development of the Junior Café Scientifique for young people in schools both in the UK and elsewhere. The informality and flexibility of the format is considered as both a strength (in terms of the movement's mission to allow participants to choose how they want to engage) and a vulnerability given the absence of an institutional funding base. Looking to the future, new models for running cafés are considered that take advantage of information and communication technologies.

Further reading and useful web sites

Each of the chapters features a selection of briefly annotated 'further reading' and 'useful web sites'. As stated above, in producing this collection we started from the premise that the enclosed readings could be considered as a starting point for students (practitioners, researchers and lecturers), but that we would also encourage further exploration of the practices of science communication through literature and associated mixed media resources. These annotated selections can be seen as initial guidance for these further explorations. Some of these have been chosen and written by the authors, others by the editing team; we would argue that they can and should be considered worthy of further study.

It should come as no surprise that several authors referred to the same sources in preparing these annotations. Rather than duplicate what are considered to be a small number of generative texts and exemplars—the online preprint archive http://arxiv.org/ being the most obvious—we have attempted to provide a selection from which students (practitioners, researchers and lecturers) can choose. It follows that we have made the decision not to mention a given 'further reading' or 'useful web site' in more than one of the chapters.

Richard Holliman,
Jeff Thomas,
Sam Smidt,
Eileen Scanlon and
Elizabeth Whitelegg

■ **REFERENCES**

Gibbons, M., Limoges, C., Nowotny, H., Schwartzman, S., Scott, P. and Trow, M. (1994).
The New Production of Knowledge; the Dynamics of Science and Research in Contemporary Societies. Sage, London.

Holliman, R., Whitelegg, E., Scanlon, E., Smidt, S. and Thomas, J. (eds) (2009). *Investigating Science Communication in the Information Age: Implications for Public Engagement and Popular Media*. Oxford University Press, Oxford.

Irwin, A. (2009). Moving forwards or in circles? Science communication and scientific governance in an age of innovation. In: *Investigating Science Communication in the Information Age: Implications for Public Engagement and Popular Media* (ed. R. Holliman, E. Whitelegg, E. Scanlon, S. Smidt and J. Thomas). Oxford University Press, Oxford.

Leadbetter, C. and Miller, P. (2004). The Pro-am Revolution: How Enthusiasts are Changing Our Economy and Society. Demos: London.

SECTION 1

Communicating post-academic science

In its post-academic mode, science can no longer evade all social responsibility by pretending that the production of universally valid, value-neutral knowledge is its only goal and only achievement.

John Ziman (2000). *Real Science*

1.1 **Scientists communicating,** *by Jane Gregory*   3

1.2 **Ethical codes and scientific norms: the role of communication in maintaining the social contract for science,** *by Robert Doubleday*   19

1.3 **Patents and the dissemination of scientific knowledge,** *by Charlotte Schulze*   35

1.1

Scientists communicating

Jane Gregory

A community in communication

This chapter gives an overview of communication among scientists. It takes a historical perspective, looking at how the development of science as knowledge, as practice and as a profession has both necessitated and made possible communication between scientists. In so doing, the chapter explores one of the fundamental properties of the scientific community. For, according to the physicist and sociologist of science John Ziman (1984, p. 58), 'the fundamental social institution of science is . . . its system of communication'.

Scientists *en masse* are most usually referred to as a community, rather than as a class or a profession. 'Community' has various meanings: it could be a group of people who share a physical location, such as the residents of a village. These people need not have anything further in common, nor need they know each other: they are a community by virtue of common location. Or, a community could be a group of people who have some particular quality in common, such as 'the Polish community in Britain' or 'the birdwatching community'. The people in these communities do not necessarily share a physical location, nor are they necessarily in touch with each other. Another definition of community is that it consists of people who hold meanings, signs and understandings in common, and share them. That is, communities consist of people who communicate with each other. They define their community by their communicative actions; it is not imposed on them by an external analyst.

There is some useful overlap between these definitions. A single birdwatcher is not the smallest possible unit of the birdwatching community: the community exists only if there are other birdwatchers with whom to share what they have in common. This thing in common might be a physical location, such as when birdwatchers gather at the site of the unusual appearance of a rare bird. But more often the 'having in common' will be expressed by participation in an organization, subscriptions to particular magazines, membership of a dedicated electronic network or devotion to particular programmes on television. That is, it is activity done by, and made possible by, the group that creates the community, and most of what the group holds in common is held there by communication.

Some theorists go so far as to say that it is the communications themselves—and not the people who generate and share them—that constitute the community. One such theorist is Niklas Luhmann, who understood communication as the exchange of meaning; and where others might think in terms of 'community' or 'society', he thought about 'social systems' (Luhmann 1995). Luhmann described social systems as being autopoietic; that is, by analogy with biological cells, they are systems that through their own action generate more of themselves. Whereas in biology this self-generation happens through feeding and reproduction, in social systems it happens through communication. Thus while scientists as individuals can be members of the community, the community itself is not the sum total of the people, as scientists do not self-replicate. In Luhmann's scheme, the self-replicating entity is communication—communications beget communications. For Luhmann, scientific communication *is* the scientific community. So while they approach the world from very different directions, Luhmann can be understood as endorsing Ziman's claim that communication is the fundamental social institution of science.

The origins of community in science

The history of science shows that communication has been central to the development of the scientific community. Historians usually date the origins of this community to the middle of the 17th century, and the formation of the first learned societies (Shapin and Schaffer 1985; Crosland 1992). Before this time, people interested in understanding the natural world usually worked in seclusion, perhaps making a living from selling craft skills or knowledge. Knowledge was esoteric, and closely guarded—passed on only as a last resort to an apprentice or heir. These custodians of natural knowledge—amateurs, alchemists, hermits and magicians—were challenged in the 17th century by a group who recruited others to a new ideology for the pursuit of knowledge. This process of recruitment was in itself a communication process, and communication was at the heart of the new philosophy—for science to be a public activity, and secrecy, private ownership and personal knowledge were outlawed. Indeed, knowledge held in secret could not, in this new ideology, possibly be science.

A public activity requires a space that is open and accessible. Some early manifestations of the scientific community were gatherings at which people could meet and talk. These gatherings were formalized as societies, which often had royal or state patronage. So these societies were public institutions of civil society. Another aspect of the new ideology of science was that nature could be accessed only through experiment (rather than transcendentally, or through pure logic, for example). But these experiments could not be done in secret, or even in private; they produced valid knowledge only when they were witnessed. So the earliest laboratories were spaces not just for experimental work but also for the witnessing of experimental work. Thus laboratories too were necessarily public; and in some of the early institutions, the lecture room and the laboratory were the same space: in the 'lecture-demonstration', science is communicated as it is done.

These founding values still inform scientific practice. In the 1930s, the sociologist and historian of science Robert Merton presented his classic study of the qualities that identify science as distinct from other professions and activities (Merton 1968). Merton noted four 'norms' of science that, for him, were definitive of the special character of science. Although these norms have been controversial in the sociology of science, they can be understood as representing an ideal which remains powerful today. Each of these norms has clear implications for communicative practice. If they are at the heart of science, then communication is too (see Doubleday, Chapter 1.2 this volume).

The first norm is communism, which expresses the public-ness of science. Good science is that which is visible, shared and held in common. So science, to deserve the name, must be communicated. The next norm is disinterestedness: science does not privilege the interests of any individual, nor is it coloured by personal preference or ideology. It is owned by, and meaningful to, the group, and so can be passed around and participated in freely. The third norm is organized scepticism, that is the exposure of new knowledge to scrutiny and critique by a disciplined community. This is clearly a communicative process. The last norm is universalism: scientific knowledge is independent of where, and by whom, it is generated. Universalism makes particular demands on communication, because it expands the scientific community beyond traditional social boundaries of culture, language and nationality.

These norms are in some ways at odds with any organizing principle. Merton locates the emergence of the scientific community in England squarely in the wake of the English Civil War (1642–52), when mechanisms for consensus-building and inclusion were being advocated for healing the nation (Sprat 1667). But communities are as much about who is left out as they are about who is included. The early scientists of the 17th century were aware of the inherent contradictions of their situation: their institutions were not accessible to all; in practice, they were an elite group. One response to this tension was to expand the community by sharing with a wider group via publications: scientific societies published journals, which included reports of meetings and scripts of lectures, and also accounts of experiments, to allow what Shapin and S. haffer (1985) call 'witnessing-at-a-distance'.

Merton's norms are expressed in the detailed as well as in the large-scale communicative practices in science. For example, disinterestedness is reflected in the impersonal, unemotional language of most contemporary scientific communication. The 'I' of the researcher is made invisible by the passive voice, and factual data take precedence over emotions and feelings (Fahnestock, 1993). These detailed factual accounts not only allow for witnessing-at-a-distance, but also enable other scientists to undertake the research for themselves. If good science is universal science, and generates the same knowledge irrespective of where or by whom it is done, then the potential for this kind of repetition of experiments must exist; and although it is rarely carried out, it is implicit in the form and content of scientific publications. Publication itself represents a triumph of organized scepticism—a norm which is routinely expressed in the process of peer review (see Wager, Chapter 4.1 this volume). Peer review involves submitting research papers to the scrutiny of colleagues who decide whether to grant scientific status to the research, and thus to permit publication. This peer-review process is also an expression of communism, in that it involves the sharing of new research among colleagues who may also be competitors; and it relies on their disinterest.

The early ambition for publications to expand the community for science was soon compromised, and publications became agents in the process of dividing the insiders from the outsiders. For example, at the Académie des Sciences in Paris, its journal at first published every submission, making editorial comment on its value as a contribution to knowledge. But very soon the Académie was unable to cope with the torrent of contributions, and it decided that some contributions would be rejected without comment. Thus rather than this journal being a public space for a wide group to share and assess knowledge as equals, it quickly became a site where insiders controlled the constitution of both their membership and their body of knowledge, through the management of communication (Crosland 1992). This 'gatekeeping' role is a significant feature of scientific publishing today, and extends, for example, to the promotion of freshly published scientific research to newsrooms (see Trench 2009).

The sociologist of science Thomas Gieryn (1995, 1999) explores this activity on the boundaries of science in order to address one of the fundamental questions in the sociology of scientific knowledge: why does society accept scientific explanations as being authoritative and reliable accounts of the natural world? Where philosophers tend to look into scientific knowledge to search for special characteristics, sociologists look at scientists and their activities to see how scientific authority is constructed. Authority is, after all, the result of a negotiation between people: it cannot only be expressed, but also has to be accepted. Gieryn identifies the social phenomena of membership (and its opposite, exclusion) and selective publication (and its opposite, rejection) as examples of a process he calls *boundary work*. Boundary work is the process by which people and ideas are identified as insiders or outsiders to a particular community. By actively rejecting that which it deems to be incompetent, irrelevant or deceptive—the 'unscientific'—the scientific community routinely, in its day-to-day business, constructs a science that has integrity, and is reliable. It affords visible badges of membership to scientists who are honest, responsible and skilled. Those of us who are outside this community understand that those within it have these attributes which we, at least in respect of scientific competence, do not have. For outsiders to accept the value of a community, the community's actions must be accessible to scrutiny. The communication that provides this access is therefore vital to the authority of science in society. It can be argued that the scientific community is one of the least policed of all the professions simply because society accepts that, in its everyday practice, science is routinely policing itself.

Communication in contemporary practice

Through four centuries of evolution and diversification, this complex of membership, practice and communication has remained fundamental to the structure of 'the scientific community'. Learned societies have multiplied; and alongside the national and royal societies there are also societies for each discipline within science, and for subdisciplines and inter-disciplines. A plethora of journals has been created by this multiplicity of societies and disciplines. Some journals have been created instead of societies, or in order to encourage and delineate an emerging discipline: readership of the journal stands in lieu of membership of a society, and the new discipline is defined by the outcome of the selection process undertaken by the gatekeepers of the new journal.

The gatekeeper for a typical journal will be its academic editor. He or she will judge the relevance of submitted papers to the subject area of the journal, and will then organize peer review. This process can take time—from weeks to years, especially if, as is common, the reviewers ask the author to revise the paper. If the paper is accepted, the subsequent publication process can also be time-consuming. One aim of the editorial process is to limit the number of accepted papers to those that the journal can publish on a reasonable timescale.

There are various types of journal; the one described here is a primary journal, that is, it publishes original research after peer review. Secondary journals review, synthesize or abstract research from the primary journals, to make it more accessible, for example to a multi-disciplinary community. Journals tend to have clearly defined subject areas, with the more specialist journals usually having smaller readerships. The more prestigious scientific journals tend to be those that are either very highly specialised or those that publish significant research in any field and have wide readerships, such as *Nature* and *Science*. With such broad scope, these journals attract a great many submissions, and some of their prestige comes from the very small proportion of submissions that are eventually published.

Scientific journals have five key functions. They are a site for peer review, which generates particular standards and directions in scientific research. They form an archive, for future reference; and register research as the scientific achievement of particular authors. They disseminate research within the scientific community and beyond; and they allow scientists to engage efficiently with their research field by identifying and locating the most significant research. Thus they serve the interests of both authors and readers.

The peer-review process works the boundary between orthodox and unorthodox science. Rejection by a research journal may be a dead end for the researcher, or suggest a return to the drawing board; but it can also be the first step towards unorthodoxy. The institutions of science reject certain instances or categories of knowledge in order to sustain a particular character for themselves. To apply Luhmann's metaphor, those which are deemed to be deficient are not allowed to self-replicate, for by so doing they would threaten the quality of the community. The deficiency may be one of competence, but it could also be one of direction—a scientist may want to pursue a goal that other scientists think is unworthy or uninteresting. Therefore the wrong kinds of science are denied access to the communications systems of science. Unorthodox scientists are literally excommunicated from the community represented by the peer-reviewed literature; and some will seek to communicate their ideas in other forums and media (Gregory 2003). Unorthodox science in popular media is therefore one symptom of the efficiency of peer review.

The full and proper presentation of research findings includes references to previous research that has informed the author's ideas and practices. These citations also identify the author's research community. They are a declaration of identity by the author—a form of scientific genealogy. Citations borrow the authority of prior research that has already received the affirmation that publication brings. Citations also identify the cited scientist as a member of a research community, and as authoritative, and therefore are a measure of esteem. So the number of citations a paper receives is used as one measure of its quality. Many journals have calculated their 'impact factor': how often is the average article cited? Any individual paper gains (or loses) prestige from the average citability of the papers published in that journal. This impact factor does not take account of the fact that a small percentage of papers generate a high percentage of citations: many papers are never cited anywhere (Chalmers, Chapter 2.2 this volume).

Scientists publishing research

In 2005, UNESCO estimated that there are five and a half million scientists worldwide. Industry figures suggest that 700,000 journal articles are published each year. This tallies roughly with other studies which suggest that the authors of peer-reviewed articles are only a small minority—perhaps 20 per cent—of all scientists (de Solla Price 1986; Bucchi 2004; Sismondo 2004; Ware 2006a). One explanation for this small number is that the traditions of communication among scientists apply very largely to academic and public-sector scientists. They do not apply to scientists in industry, who may serve different interests and values, and are rewarded in different ways. Scientists in industry may be communicating within their own commercial environment rather than with their competitors; and intellectual property documentation serves many of the same functions, though in a legal framework, of claim-staking and archiving that journals serve for academic scientists (see Schulze, Chapter 1.3 this volume). In the United States, between 10 and 20 per cent of all scientists work in universities rather than in the commercial sector, and yet they publish 75 per cent of papers in research journals (Ware 2006a). Another explanation for the low percentage of scientists who participate in what are represented as the traditional and routine communication practices of the scientific community is that it is an elite community with very effective self-advertising, and it maintains the authority of the whole through the activities of the few. Thus those scientists who publish in the peer-reviewed literature earn the respect that society grants to science as a whole; and the reliable knowledge they generate there confers reliability on scientific knowledge in general.

Scientists communicate not only in print but also in person. Meetings among large groups of scientists are usually organized as conferences, and the formal proceedings consist of the presentation of recent research. Like the publication process, participation in a conference is often the result of a selection process, and therefore the mere fact of participation can be taken as an achievement. On a scientist's CV, this participation will have more value than any reaction to the presentation, which will most likely go unrecorded.

A relatively recent form of communication among scientists relates to the funding arrangements for science. In order to gain funding from public authorities, charities or trusts, scientists are now required to write detailed accounts of the research that they wish to undertake. Funding applications must set out not only research aims and methods, but also a detailed budget and the projected outcome of the research in terms of its scientific significance, its potential applications and the intended 'end-users' of the research. In recent years, applicants have also been asked to outline their strategy for communicating their research outcomes, and to budget for this.

These funding applications are enormously time-consuming for scientists to prepare; indeed, some of the better-resourced scientific institutions now employ dedicated officers who work with scientists on their applications. This has become such a significant part of a scientist's daily work that it could reasonably be added to the traditional conference talks and journal papers as a core communications activity. However, it differs from these other communications in that the application document may be seen by very few people—a couple of administrators, and a couple of reviewers—all of whom are bound to uphold the confidentiality of the process; and there is rarely much opportunity to amend and revise as would be routine for a peer-reviewed paper. Since the majority of funding applications are never funded, the ideas and efforts that they embody will, by and large, never be appreciated by the community. Raising funds to do research is an important professional skill in science, and successful funding applications, and the money they raise, will be listed on scientists' CVs. However, anecdotal evidence now suggests that scientists, begrudging the effort they invest in unsuccessful applications, are also listing these failures as professional achievements on their CV.

Since the 1960s the productivity of scientists and their institutions has been measured in terms of quantity of communication. This measuring, which is called bibliometrics or scientometrics, counts units of publication as an indicator of scientific quality and achievement. Similarly, citations, by other scientists, of published papers have been counted to measure the value of individual authors, or of individual publications. Although bibliometrics is considered by sociologists to be a rather blunt instrument (Bucchi 2004; Sismondo 2004), it is still used, for example, in the UK government's Research Assessment Exercise, to judge the performance of universities. In 2007, British university scientists (and other academics) were asked to declare their four most prestigious publications from a given period of assessment, and to justify their claim to prestige in terms of originality of the work, the status of each publication and its international reach. This kind of assessment takes no account of the audience for, or reception of, these communications: given the organization and practice of the scientific community, the mere fact that these publications exist is taken as a significant indicator of the status of their author. Although these metrics are subject to change for political reasons, when they stabilize they become organizing frameworks for scientific activity, and scientists comply to achieve high scores.

The sociologists of science who moved on from bibliometrics did so for many reasons. One is its emphasis on quantity rather than quality. But another was a growing understanding of the role of informal communication in science. The scientists' community is enacted not just through journals but also through chats in the corridor, or by e-mail. At a conference, the most valuable communication may take place not in the lecture hall but over dinner. While it can be easy to categorize a communication as formal or informal, it is less easy to see any clear advantage of one over the other as contributing to scientific knowledge: the intellectual and practical activity that is science is happening in both places (Collins and Pinch 1979). This perspective has prompted many 'ethnographic' studies which involve direct and detailed observation and recording of scientists' everyday working life (Latour and Woolgar 1979; Lynch 1985; Sismondo 2004).

It is tempting to ask whether most, if not all, science is done by informal rather than formal communication. Formal communication would then offer legitimation (Sismondo

2004), and serve as public relations, and as something to measure. So a formal account might be constructed *post hoc*, in order to legitimate knowledge developed through informal means (Gregory 2003). The differences one can observe between the accounts of research given in journal papers and what one might see happening in an ethnographic study of the same research certainly lend some credence to this view. It may also account for the small percentage of scientists who publish in the peer-reviewed literature: the others are busy doing science via informal communication.

Another blurred boundary exists between communication among scientists, and communication between scientists and publics. A significant consensus has emerged around the idea that an important audience for popular science is scientists (Gregory and Miller 1998). Many studies have shown how scientists use popular science in their professional work, where it functions as professional communication (Shinn and Whitley 1985; Hilgartner 1990; Gregory 2005). This is especially so in new subject areas, and in interdisciplinary fields. Scientists also fulfil professional functions in many of the contexts in which they are required to communicate in the public sphere. For example, when scientists act as policy advisers to government, as consultants to industry or as expert witnesses in a court of law, they are communicating with non-scientists, but they do not cease to be scientists in the process. Indeed, their communications would be redundant if they did.

Scientists have also, in recent decades, been encouraged to communicate via mass media and with publics more directly (face-to-face and via new media), as part of their professional duties (see Holliman *et al.* 2009). Over the same period, relationships between the primary scientific literature and the news media have been transformed, with most of the major journals now press-releasing selected research papers (Trench 2009). This means that the rewards of being published in a major journal are not just the prestige and measurability that formal publication brings a scientist, but potentially also media attention. It has even been suggested that editorial selection and peer review now take account of the potential of the research as a news story (Kiernan 2006; Bauer and Gregory 2007). If this is the case, then it will be interesting to see how the emphasis on media professionals' attribution of news values will affect the composition of the primary scientific literature, and so the direction of science itself.

The trend in science is not all in more public directions. Since the 1970s, scientific research in the UK has been increasingly funded by corporations and private capital, either outside the universities or in partnerships. In this 'post-academic' science (Ziman 2000), knowledge is owned not by its discoverer or by the scientific community but by the corporation. Some scientists retain ownership of their knowledge by becoming a corporation: scientists 'starting up' small businesses, within or outside universities, are now common. This tends to happen where there is an immediate prospect of marketable products, and these products usually have to compete in the marketplace. This competitive aspect frames the communication about the science: the corporation retains ownership by communicating in the intellectual property literature, rather than in the academic journals.

The 'small business' model has also been adopted for organizing publicly funded research in universities, with scientists working as 'firms' that may be delineated differently from their usual discipline or department. This is one of the reasons why there are now many more scientists working in teams, and why most journal papers in science are written by more than one author. The size of these 'publishing teams' increased through the 20th century, with the number of papers by only one author halving between 1920 and 1950. The trend for co-authorship is still rising: in 2003, the average number of authors per paper was 4.8, compared to 2.3 in 1988.

While scientists are publishing more papers as co-authors, these fractional contributions add up to fewer 'whole papers': in 1950, a scientist would publish one paper per year (or its equivalent in fractional contributions); in 2000 the average was 0.7. So scientists now typically have their names on more papers, but their fractional contributions add up to less research (Ware 2006a). This suggests that productivity measures based on counting publications may have actually decreased scientific productivity.

International research teams are common. From the earliest era of modern science, international communication has been very important. The use of Latin as the international language of scholarship facilitated this exchange in Europe. Travel and hospitality were important for scientists looking to build a career, which gave diplomats, soldiers and merchants a head start, and made trading cities important loci of scientific activity. For example, the huge effort of theory and classification in natural history in the 18th and 19th centuries took place in the international framework of expanding empires, where civil servants, missionaries and sailors did the scientific work of gathering materials and observations inaccessible to their contacts in the European capitals, and they carried with them books and instruments.

From the 18th century onwards scholarship was most often conducted in the vernacular, and scientists wishing to achieve an international career would learn foreign languages. But in the latter half of the 20th century the changing political landscape of the West and the success of American science pushed English into the lead as the international language of the scientific community, and English-language training is now part of a scientific education in many countries. International communication, along with travel, remains important in a scientific career. Publication in an international journal tends to be ranked more highly than publication in a local journal, and a paper presented at an international conference is deemed more significant than a paper at a national one. International collaboration is often encouraged by national policies and enforced by the criteria set by funding agencies, and has often been used as a tool of foreign policy. While political rivalries have brought tensions into international scientific cooperation, the proclaimed disinterestedness of scientists has provided them with some immunity, and enabled them to serve as informal ambassadors, for example in post-war reconstruction, or to prepare the ground for trade agreements.

The business of scientific communication

While scientific journals have their origins in the meetings of scientific societies, commercial publishers have been involved in producing them since the 18th century (Brock and Meadows 1983), and now publish two-thirds of all research papers. Publishers have been key players in the founding of disciplines, by trusting that a new research programme is likely to generate a community that will make a journal financially viable. In 2006 it was estimated that there were 2000 journal publishing companies around the world, of which about 650 were publishing journals in English. Of these 650 companies, 70 per cent were 'not for profit', and produced 20 per cent of journals. A small number of companies dominate the industry, with the top 11 (2 per cent) producing more than 70 per cent of journals (Ware 2006a).

Scientific publishing is big business. Annual revenues are estimated at $9–12 billion, of which about half is from the English-language market. This market is a relatively small part of the global publishing industry (about 5 per cent) but the production of scientific journals employs around 100,000 people, of whom about half are in the EU. They produce around 23,000 scholarly journals, and 1.4 million articles, per year (Ware 2006a). The number of peer-reviewed journals has grown at a steady rate of 3.5 per cent per year for over two centuries. This reflects the expansion of the professional community for science, and the proliferation of new disciplines (Bucchi 2004). The number of articles catalogued increased by 50 per cent between 1988 and 2003, with the most significant contribution to this growth coming from East Asia.

During the 1990s, journals became available online, and by 2003 two-thirds of journals world-wide, and almost all of those in English, could be accessed by computer by members of subscribing institutions (Ware 2006a). This development has led to greatly increased potential access to the scientific literature, and a greatly reduced cost per use (the marginal costs of electronic distribution are very low). There are a number of schemes for providing free or discounted online access to the scientific literature for scientists in developing countries.

Around 10 per cent of journals are currently (2008) available online without subscription. These are called open-access journals. There are two main forms of open access: one in which the journal's publisher makes the articles available to open access, and the other in which the author chooses to make his or her article available in this way, usually after publication on paper. Some authors undertake this 'self-archiving' in the expectation that the greater availability of their paper will mean more citations, though whether this happens has been debated (Lawrence 2001; Eysenbach 2006). In general, physicists and computer scientists are among the most enthusiastic users of online archives, whereas interest among other disciplines is much lower.

Open-access is funded in a variety of ways, but the most common is that the author, or his or her institution, pays a fee if the paper is published. Rejected authors pay nothing, even though the peer-review process incurs real costs. So far open-access publishing is generally not breaking even, and some has public funding. For example, Cornell University has run arXiv (X for the Greek 'chi') since 1991 with funding from the US National Science Foundation (see Chalmers, Chapter 2.2 this volume for further discussion).

Around half of publishers are happy for authors to self-archive their papers online, and only one-fifth explicitly forbids it. However, only 15 per cent allow online self-archiving before publication, and some have embargo periods with a view to protecting subscriptions. Publishers are concerned about the impact of widespread self-archiving of journal articles, since if the papers are free to anyone via this route there is less reason for libraries to pay for the journals (Ware 2006b). Indeed, the financial relationships between libraries and publishers have been turbulent and controversial during the implementation of online services (Hitchcock *et al.* 1997). In the UK, there has been tension within government between departments promoting education and science who are looking to increase online services, and those looking to protect the financial interests of the traditional publishing industry (House of Commons Select Committee on Science and Technology, 2004).

New communications technologies

Online journals use the computer screen as a cheap and accessible substitute for paper; however, novel forms of scholarly communication are emerging from newer network technologies. Scientists were the first users of electronic networks: in 1965 the US Department of Defense developed a network among universities in the western states of the USA to enable scientists to share a small number of fast computers (Slevin 2000). This 'internet' grew and spread, and with the arrival of e-mail in the 1970s scientists used it to collaborate on research projects and share their results. The leakiness and vulnerability of the internet is a legacy of its early years as a medium for scientists (Agre 1998): the scientists' strong norms, and the sense of community, made security a low priority—rather like a village where no-one feels the need to lock their door.

Although the invention of the World Wide Web at CERN in 1989 was aimed at enabling scientists in dispersed communities to work together (Slevin 2000), scientists have been slower than other groups to take advantage of network technologies. Wikis, blogs and user-networks have proliferated in many spheres (see for example Chalmers, Chapter 2.2 this volume) but it seems that scientists in general are engaging with these technologies only very conservatively: for example, in 2005, *Nature* reported that of 20 million blogs, only a few dozen were among scientists (Butler 2005). However, some fields—string theory is one—have used blogs as forums in which colleagues can share ideas and respond to suggestions, thus extending the sphere for the informal communication via which scientific work can be done. Wiki tools, which enable contributors to amend and develop material that has already been posted, have enabled the ongoing development of common resources, for example in protein sequencing where scientists around the world are contributing to a larger project.

Scientists have mixed feelings on the use of blogs as a substitute for the formal communication that is found in journals: for some, the blog is a fast and democratic equivalent, with the whole community acting in the role of journal editor; but for others, this sharing of responsibility is one for which they have little time and which brings little reward. It is difficult to see how scientific achievements made through blogging could be connected with particular authors. Does anyone who joins a blog or amends a wiki become a co-author of any published outcome (peer reviewers, who may insist on changes to a paper, but do not make them themselves, do not)? Could someone watching a blog write up the ideas and publish them on paper, trumping the bloggers?

One concern about the democratic potential of blogs is that, by being open to everyone, they are also accessible to people who are not specialists, or who might take the discussion off at a tangent, or are simply cranks. For some scientists, this is a reason for avoiding these tools completely: any space that accommodates unorthodox science is no place for real science. So blogs may have to compromise, and appoint moderators to maintain quality and focus. But there is a more positive attitude towards blogs as a form of informal communication, as Chalmers argues in this volume (Chapter 2.2): they can be useful for constructive dialogue with dissenters without compromising the exclusivity of formal publications, and for discussion of broader issues—climate scientists can get into politics, and evolutionary biologists into creationism. And unlike many forms of informal communication, blogs are easy to archive and search.

So blogs are valued for their hybrid quality: poised somewhere between the formal and the informal, they can be a useful space for research that doesn't quite fit into traditional categories, such as research outputs that are not big enough (or quite ready) for a whole, formal paper but are still worth sharing; or which take more risks with data than would survive peer review. Some user networks are now being developed in collaboration between publishing companies and research communities, often with funding from advertising, which combine raw data sets, informal communication and peer-reviewed publication. One such is the BIOSCI/BIONET (http://www.bio.net/docs/biosci-termsofuse.html), run by the University of Indiana, which is currently (2008) a free service that 'promotes communication between professionals in the biological sciences.' The site 'reflects and facilitates the scientific community's passion for public discourse to arrive at truths in an open manner'. No fees are charged, and messages are distributed without editorial intervention, and across the globe.

However, the general feeling seems to be that these tools are, at the moment, just one more thing that busy scientists could do, and for little reward. If and when they become an accepted substitute for part of the present workload, and there is a clear system of credit for contributors, this attitude may change. Some commentators note a 'generation gap' in attitudes to blogging, and think that it is gaining ground with younger scientists who will be leaders in the future. But for the most part, and for now, it appears that as long as the journal paper is the dominant unit of credit, it will continue to dominate the hierarchy of communication.

Is there a global scientific community?

Globalization was a strong theme in intellectual and popular culture of the late 20th century (Lyotard 1984; Giddens 1990; Beck 1992). Scholars identified very large-scale systems of organization and interaction, such as the system of trade, in which raw materials, manufacturing, and markets are routinely far apart, and connected by long-distance transportation, language translation and international currency exchange. The global system of mass communication is held together by capital, technology and audiences, and information technology has been advocated as a largely benign globalizing force. The long-standing internationalism of scientists and their professional norms makes them ideal citizens in a world that is operating more and more on a global scale; indeed, the community of scientists is now often referred to as the global scientific community.

But as with other aspects of globalization, the idea of a global community of science is problematic. This community is one of common location, but only because Planet Earth is its only option. Like the Polish community in Britain, its members have a common heritage and customs; and like the birdwatchers, they share some interests and practices. But is it a community in communication? In terms of the various fields of science, it is clear that there are many which are entirely unconnected. The proliferation of specialist journals suggests smaller, more dispersed communities, rather than larger ones, and they speak distinct technical languages (Montgomery 2004). There is clear evidence that scientists can easily increase their publication counts by publishing identical research in different disciplinary or linguistic communities (Mounir and Garner 2008). That this 'self-plagiarism' can occur when there is such a strong ethos in science of publishing only original material suggests that communication among wider groups is poor.

Given the growing proportions of scientists in business rather than in academia, and the relatively low participation of commercial scientists in the traditional communications of the scientific community, there may be another divide: between academia and business. The scientific community in business may itself be fragmented, given that its communications are more about competition and self-interest than they are about Mertonian universalism and communism (Bauer and Gregory 2007).

The growth of journal publishing in East Asia—along with the economic boom in the region—suggests that we may anticipate another evolutionary jump in the international language of science. Commentators have identified Mandarin as the next international language of business, though they do not anticipate the same turn in science; but it is possible that two parallel linguistic communities could emerge, and the successful scientist would be the one who can communicate in both. If online technologies become significant, they may bring with them further divisions: they may be more available and technically reliable in some regions than in others, and their pattern of use in other spheres, where demographic inequalities are clear, may be repeated in science too. There are already hints of further fragmentation within the online tools themselves, where discussions divide into threads to maintain focus and keep traffic to manageable levels.

In a globalized world, therefore, the very idea of 'the scientific community' may be a fiction. If so, it remains a powerful one. The scientific community, idealized in Merton's norms, and transacting its boundary work to display its integrity, has become a persistent authority in society. Traditional communicative practice is essential not just to the display of this authority, but to the construction of authoritativeness. It is perhaps therefore not so surprising that the formal, peer-reviewed journal remains so central to accounts of scientific practice (including this one), despite the fact that only a small percentage of scientists ever publish in one. And it is also not so surprising that scientists are reluctant to abandon the traditions that grant them authority, despite the lure of emerging forms of (mainly electronic) communication; or that the scientists most keen to innovate in communicative practice are those who, like the string theorists, are rather distant from the sites of contested authority in society. When science is called into question, scientists often insist that the peer-reviewed literature is the only reliable source of knowledge. For scientists whose knowledge could potentially poison or pollute, the peer-review system is their safety net and their licence, and society accepts it as such. While this culture persists, so will the peer-reviewed journal.

■ REFERENCES

Agre, P. (1998). The Internet and public discourse. *First Monday*. Available online at: http://www.Firstmonday.dk/issues/issue3_3/agre/.

Bauer, M. and Gregory, J. (2007). From journalism to corporate communication in post-war Britain. In: *Journalism, Science and Society: Science Communication Between News and Public Relations* (ed. M. Bauer and M. Bucchi), pp. 33–52. Routledge, London.

Beck, U. (1992). *Risk Society: Towards a New Modernity*. Sage, London.

Brock, W., and Meadows, A.J. (1983). *The Lamp of Learning*. Taylor and Francis, London.

Bucchi, M. (2004). *Science in Society: an Introduction to Social Studies of Science*. Routledge, London.

Butler, D. (2005). Science in the web age: joint efforts. *Nature*, **438**, 548–9.

Collins, H. and Pinch, T. (1979). The construction of the paranormal: nothing unscientific is happening. In: *On the Margins of Science: the Social Construction of Rejected Knowledge*, Sociological Review Monograph No. 27 (ed. R. Wallis), pp. 237–70. University of Keele Press, Keele.

Crosland, M. (1992). *Science Under Control: the French Academy of Sciences 1795–1914*. Cambridge University Press, Cambridge.

Eysenbach, G. (2006). Citation advantage of open access articles. *PLoS Biology* **4**(5), e157.

Fahnestock, J. (1993). Accommodating science: the rhetorical life of scientific facts. In: *The Literature of Science – Perspectives on Popular Scientific Writing* (ed. W. McRae), pp. 17–36. University of Georgia Press, Athens, GA.

Giddens, A. (1990). *The Consequences of Modernity*. Polity Press, Cambridge.

Gieryn, T. (1995). Boundaries of science. In: *Handbook of Science and Technology Studies* (ed. S. Jasanoff, G.E. Markle, J.C. Petersen and T. Pinch), pp. 393–443. Sage, Thousand Oaks, CA.

Gieryn, T. (1999). *Cultural Boundaries of Science: Credibility on the Line*. Chicago University Press, Chicago, IL.

Gregory, J. (2003). Popularisation and excommunication of Fred Hoyle's 'life-from-space' theory. *Public Understanding of Science*, **12**, 25–46.

Gregory, J. (2005). *Fred Hoyle's Universe*. Oxford University Press, Oxford.

Gregory, J. and Miller, S. (1998). *Science in Public: Communication, Culture, and Credibility*. Plenum Press, New York.

Hilgartner, S. (1990). The dominant view of popularization: conceptual problems, political uses. *Social Studies of Science*, **20**, 519–39.

Hitchcock, S., Carr, L. and Hall, W. (1997). Web journals publishing: a UK perspective. *Serials*, **10**(3), 285–99.

Holliman, R., Whitelegg, E., Scanlon, E., Smidt, S. and Thomas, J. (eds) (2009). *Investigating Science Communication in the Information Age: Implications for Public Engagement and Popular Media*. Oxford University Press, Oxford.

House of Commons Select Committee on Science and Technology (2004). *Tenth Report* [inquiry into scientific publications]. HMSO, London. Available online at: http://www.publications.parliament.uk/pa/cm200304/cmselect/cmsctech/399/39902.htm.

Kiernan, V. (2006). *Embargoed Science*. University of Illinois Press, Champaign, IL.

Latour, B. and Woolgar, S. (1979). *Laboratory Life: the Construction of Scientific Facts*. Princeton University Press, Princeton, NJ.

Lawrence, S. (2001). Free online availability substantially increases a paper's impact. *Nature*, **411**, 521.

Luhmann, N. (1995). *Social Systems*. Stanford University Press, Stanford, CA.

Lynch, M. (1985). *Art and Artifact in Laboratory Science: a Study of Shop Work and Shop Talk in a Research Laboratory*. Routledge and Keegan Paul, London.

Lyotard, J.-F. (1984). *The Postmodern Condition: a Report on Knowledge*. Manchester University Press, Manchester.

Merton, R. (1968). *Social Theory and Social Structure*. Free Press, New York.

Montgomery, S. (2004). Of towers, walls and fields: perspectives on language in science. *Science*, **303**, 1333.

Mounir, E. and Garner, H. (2008). A tale of two citations. *Nature* **451**, 397–9.

Shapin, S. and Schaffer, S. (1985). *Leviathan and the Air Pump: Hobbes, Boyle and the Experimental Life*. Princeton University Press, Princeton, NJ.

Shinn, T. and Whitley, R. (1985). *Expository Science: Forms and Functions of Popularisation*. Reidel, Dordrecht.

Sismondo, S. (2004). *An Introduction to Science and Technology Studies*. Blackwell, Malden, MA.

Slevin, J. (2000) *The Internet and Society*. Polity Press, Cambridge.

de Solla Price, D. (1986). *Little Science, Big Science and Beyond*. Columbia University Press, New York.

Sprat, T. (1667). *The History of the Royal Society of London for Improving of Natural Knowledge*. London.

Trench, B. (2009). Science reporting in the electronic embrace of the Internet. In: *Investigating Science Communication in the Information Age: Implications for Public Engagement and Popular Media* (ed. R. Holliman, E. Whitelegg, E. Scanlon, S. Smidt and J. Thomas). Oxford University Press, Oxford.

Ware, M. (2006a). *Scientific Publishing in Transition, an Overview of Current Developments*. Mark Ware Consulting Ltd, Bristol. Available at: **http://mrkwr.wordpress.com/articles/**.

Ware, M. (2006b). Open archives and their impact on journal cancellations. *Learned Publishing* **19**(3), 226–9. Available at: **http://mrkwr.wordpress.com/articles/**.

Ziman, J. (1984). *An Introduction to Science Studies: the Philosophical and Social Aspects of Science and Technology*. Cambridge University Press, Cambridge.

Ziman, J. (2000). *Real Science*. Cambridge University Press, Cambridge.

■ FURTHER READING

- Bucchi, M. (2004). *Science in Society: an Introduction to Social Studies of Science*. Routledge, London. The first half of this book introduces readers to a number of sociologically informed perspectives in the study of science in society. In the second half of the book Bucchi examines a series of high-profile science-based controversies, including the 'Sokal affair' (also known as the 'science wars'), global warming, genetically modified foods and the human genome project.

- Montgomery, S. (2003). *The Chicago Guide to Communicating Science*. University of Chicago Press, Chicago. In this wide-ranging book Montgomery offers practical advice on how to communicate science through a range of forms, including grant proposals, research papers, conference presentations and web-based and popular media.

- Slevin, J. (2000) *The Internet and Society*. Polity Press, Cambridge. Although not specifically focusing on science, this book provides a theoretically informed account of the impact of the internet on modern culture and communication. In so doing, the author relates these theoretical arguments to substantive examples of internet use from around the world.

- Ziman, J. (2000). *Real Science*. Cambridge University Press, Cambridge. Ziman's text argues—through examination of the practices of contemporary science, both in terms of how scientists conduct and communicate their research—that the culture of 'academic' science has changed to a 'post-academic' mode. In so doing, he draws on his considerable academic experience, both as an eminent condensed-matter physicist and scholar of science, technology and society.

◼ USEFUL WEB SITES

- **Science Commons: http://sciencecommons.org**. Science Commons is described on its web site as designing: '. . . strategies and tools for faster, more efficient web-enabled scientific research. We identify unnecessary barriers to research, craft policy and contracts to lower those barriers, and develop technology to make research data and materials easy to find and use'. The site has an associated blog.

1.2

Ethical codes and scientific norms: the role of communication in maintaining the social contract for science

Robert Doubleday

Introduction

What scientific research should be paid for with public funds? What authority should scientific knowledge carry in society? These questions are being asked with renewed intensity at a time when the demands on science seem to grow inexorably. From global climate change to economic competitiveness, from obesity to the supply of clean water, policy-makers and publics are calling on science to understand, and help solve, a wide range of complex problems. And yet, the legitimacy of scientific expertise is routinely challenged (see for example Irwin and Wynne, 1996). In the context of political controversies, scientists are recognized by publics as having vested interests in the claims they make, and new technologies are not seen as unquestionably advancing the cause of human progress.

The dilemma of understanding the place of science in society is not new. In a recent book on the politics of biotechnology, Jasanoff (2005, Chapter 9) summarizes the contemporary debates about science in terms of a renegotiation of a new 'social contract' for science. The social contract describes the institutional arrangements for science put in place in the USA and many other western countries following the World War II (i.e. post-1945). At this time the social prestige of science was at a high point following the demonstration of its utility during the war effort. Science policy-makers like Vannevar Bush, director of the US Office of Scientific Research and Development, translated the high status of science into an institutional mechanism designed to ensure both public funding of research and the autonomy of the scientific community in deciding how this money should be spent. In *Science: the Endless Frontier* Bush (1945) recommended to the then US president a blueprint for the setting up of the National Science Foundation (NSF) at arm's length from government with the aim of promoting 'basic research'. In return for the freedom and resources granted to science, society would receive scientific knowledge, technologies and scientifically trained labour. Under the implicit terms of this social contract it was considered inevitable that the products of science would be in the interest of society at large. Thus the state provided funds and guarantees of autonomy; science used public funds in the pursuit of knowledge; industry translated scientific knowledge into useful products; and the public supported science through taxation, enjoyed the fruits of science-driven innovation and deferred to the expertise of scientists.

This social contract for science has, however, proved vulnerable to questions about the public value of the science and technology it produced. During the decades that followed the establishment of the post-war social contract, science and technology became associated not only with securing military victory but also with the devastating effects of war; with ecological harm as well as human progress; and then, during the economic crises of the 1970s, with failing to deliver economic benefits through innovation. Jasanoff (2005) argues that this questioning of the social contract took different paths in different countries. However, a common thread to this renegotiation of the social contract was an attempt to direct scientific research towards the generation of social benefits. For example, in the USA the passing of new legislation in 1980 encouraged the commercialization of research through patenting.[1] In Europe, some analysts argued that the potential applications of scientific knowledge should not be an afterthought, but carefully considered during the design and conduct of research. Rather than disciplinary (mode 1) research, Gibbons *et al.* (1994) argued that interdisciplinary, application-oriented (mode 2) research was becoming increasingly important.

Such discussion of the 'social contract' for science highlights the conditional relationship between science and its social and political context. Science, like any other large-scale social activity, requires a supply of money, labour and political support to continue operating. As the external political and economic conditions change it should be expected that the social contract for science will also shift.

In this chapter I explore two attempts to address the question of the relations between science and the wider society in which it operates. The first episode is a contemporary example, taking place in the UK during the first decade of the 21st century; the second historical episode took place in the USA in the late 1930s and early 1940s. Both concern efforts to shore up the prestige of science against what the authors of these initiatives saw as threats from 'anti-science' tendencies in society. By focusing on the dimension of communication, I highlight the strengths and limitations of these two approaches in providing an account of science that helps us to understand its evolving relations with its social and political context.

---

1. The Bayh–Dole Act of 1980 encouraged universities and academics to patent research funded by the US federal government, for a brief overview see Kennedy (2005; also Schulze, Chapter 1.3 this volume).

This chapter is not concerned with explaining the causes of the current renegotiation in the social contract for science.[2] Rather, this chapter seeks to contribute to understanding the current place of science in society in the UK by focusing on questions of science communication. In particular, it looks at how sociological and ethical accounts of the workings of science conceptualize communication among scientists and between the scientific community and wider society.

This chapter begins therefore by looking at a recent high-profile effort in the UK to rebuild public trust in science through formalizing an ethical code describing how scientists should treat each other, their research subjects and the society and natural environment in which they work. This code was drafted and promoted by Professor Sir David King, then the UK government's chief scientist. It aims to be universal in the sense that it is designed for scientists working in academia, government and industry anywhere in the world. In order to highlight what is at stake in this ethical code, and what assumptions it makes about science communication, I compare it to an intervention made 60 years ago by the American sociologist Robert Merton. In this case he was seeking to defend science not from public mistrust but from domination by totalitarian states, such as those of fascist Germany and communist USSR.

In the next two sections I introduce King's ethical code and then Merton's description of the scientific ethos. Following this discussion of the two characterizations of the moral order of science, I outline four modes of science communication that can be traced through these two descriptions. The first two modes relate to scientists' communication among themselves: constitutive communication necessary for the production of scientific knowledge; and strategic communication that allows scientists to make decisions about their research trajectories. The second two modes of communication are between scientists and society external to science: dissemination of scientific knowledge as a 'finished product' from science to its publics; and communication that has as its primary aim the building of public trust in science through dialogue. Finally, in the conclusion, I cast further doubt upon the feasibility of King's ethical code by briefly considering a current area of major scientific research and development where trust may well prove hard to ensure—that of nanotechnology.

King's 'Universal ethical code for scientists'

From 2000 to 2007 Professor Sir David King was the UK government's chief scientific advisor. At an international meeting of science ministers and advisers in 2004 King proposed the development of an ethical code that could be voluntarily adopted by scientists and scientific institutions across disciplinary and national boundaries. The text of the code was then written by King and a small group of British academics. In May

---

2. A variety of explanations have been offered for the crisis in the social contract, including pressures from unmanageable technological risks (Beck 1992); a decline of public trust in expertise (House of Lords 2000); a wider change in the political arrangements that led to the social contract (Nowotny *et al.* 2001); or a combination of these reasons (for a critical discussion of public attitudes to expertise see Wynne 1996).

2005 the code was published by the UK government's science policy advisory body, the Council for Science and Technology, which asked for comments on how the code could be put to use (Peters 2005).

Over the following 2 years the ethical code was adopted on a trial basis by scientific agencies of the British government.[3] In September 2007 the Department for Innovation, Universities and Skills (DIUS) published the code in the form of a leaflet designed for wide circulation. That same month King used the occasion of the annual festival of the British Association for the Advancement of Science to call on individual scientists and scientific bodies to adopt the code. In brief, the code urges scientists not to cheat or mislead; to work carefully and acknowledge the contribution of others; to minimize harm to people, animals and the environment; and discuss and reflect on the issues science raises for society (Box 1). But for my argument, it is not just the content of the code that is of interest, but also the reasons King gave for its introduction. The publication and promulgation of a universal ethical code is explicitly designed to engender public trust in science.

BOX 1 RIGOUR, RESPECT AND RESPONSIBILITY: A UNIVERSAL ETHICAL CODE FOR SCIENTISTS (DIUS 2007)

Rigour, honesty and integrity

- Act with skill and care in all scientific work. Maintain up-to-date skills and assist their development in others.
- Take steps to prevent corrupt practices and professional misconduct. Declare conflicts of interest.
- Be alert to the ways in which research derives from and affects the work of other people, and respect the rights and reputations of others.

Respect for life, the law and the public good

- Ensure that your work is lawful and justified.
- Minimize and justify any adverse effect your work may have on people, animals and the natural environment.

Responsible communication: listening and informing

- Seek to discuss the issues that science raises for society. Listen to the aspirations and concerns of others.
- Do not knowingly mislead, or allow others to be misled, about scientific matters. Present and review scientific evidence, theory or interpretation honestly and accurately.

---

3. The agencies that trialled use of the code were the Environment Agency, the Veterinary Laboratories Agency, the Forestry Commission, the Defence Science and Technology Laboratory and the Pesticide Safety Directorate. There is little published information on how the code was used or how the trial was evaluated, but according to the government's own account: 'Results were positive, and showed clearly that the code was applicable in practical settings' (http://www.berr.gov.uk/dius/science/science-and-society/public_engagement/code/page41296.html. However, anecdotal evidence suggests that its impact 'on the ground' to date (2008) for those not involved in these trials is modest.

The stated aims of the code are 'to foster ethical research; to encourage active reflection among scientists on the implications and impacts of their work; to support communication between scientists and the public on complex and challenging issues' (DIUS 2007). However, in comments made by David King at the British Association Festival in September 2007 and in an article he wrote for *The Guardian* (King 2007) his focus was on renewing public trust in science. For example, he argued in a *Telegraph* article that: 'We want our new developments, treatments and technologies to be trusted and used. That's why I have championed a universal ethical code of scientists to help us to further build an environment where science and scientists are recognised as a valued part of society' (King, quoted in Highfield and Fleming 2007).

The publication of the code is intended to address what many science policy-makers consider a potentially pathological level of public mistrust in science. For King (2007) this public mistrust is evident in the controversy over genetically modified crops and the drop in vaccination rates in the UK following speculation about a link between the combined measles, mumps and rubella (MMR) vaccine and autism. In his article, King suggests that a cause of public mistrust is scientific fraud. King specifically mentions the case of the Korean stem-cell scientist Hwang Woo-suk, who in 2006 was forced to resign his university post after he was found to have falsified the results of cloning experiments. Thus King implies that the small number of high-profile cases of scientific fraud are a significant cause of public mistrust. King argues that publicizing the ethical nature of science will restore public confidence in science as it will 'demonstrate to the public that scientists take ethical issues seriously' (King 2007).

It is possible to summarize King's position by highlighting the three elements of his argument. First, the success of science depends upon public support. Second, public support for science depends on ensuring the ethical conduct of science, and communicating this ethical commitment to the wider public. Finally, that the scientific method itself is the principal mechanism for ensuring the ethical conduct of science in the public interest, albeit augmented by active reflection on the effects of research on research subjects, the environment and society.

Merton's 'Normative structure of science'

David King's concern with making explicit the ethical code of science at a time of realignment in the relations between science and wider society has striking echoes with a period of debate in the 1930s. At that point the relationship between science and the state was discussed in terms of the degree to which science should be free to set its own research agendas.

The forerunner of the 'social contract' for science promoted by Vannevar Bush in the USA was the UK's 'Haldane principle of research council autonomy'. During the second decade of the 20th century R.B.S. Haldane, an eminent politician, argued that research carried out for government departments could be divided between that which was directly applicable to the work of the commissioning department and that which was more generally useful. Haldane proposed that the more general research should be administered by autonomous research councils, and in 1920 the Medical Research Council was formed (Duffy 1986). However, not everyone accepted this emergent divide between 'applied' science, which could be more directly steered by the state, and the higher status 'basic' science, which was the autonomous domain of scientists. During the 1930s an influential group of left-wing scientists in the UK formed the 'social relations of science' movement, which argued that the state should play a more active role in planning scientific research in order to achieve more socially desirable outcomes (McGucken 1984). This debate between the planners and those supporting the autonomy of science was the backdrop for news of the adoption of an explicitly racist science policy by Nazi Germany.

In response to reports of Jewish scientists being fired from German universities, Robert Merton, the American sociologist, published two short papers (Merton 1938, 1942) about the unwritten moral codes that operate within science. He argued that there is a specific set of social norms that regulate how scientists behave with respect to their work and each other, and the operation of these norms supports the scientific method and advances the goal of knowledge production. Merton was spurred to make this argument by reports of Nazi Germany's policy of 'racially purifying' science, but his argument was that totalitarianism in any form is detrimental to the interests of science because it disrupts the autonomous social order of the scientific community.

In 'Science and the social order' and 'Science and technology in a democratic order' published in 1938 and 1942, respectively, Merton argued that science had developed as a social institution with a particular internal moral order that functions to produce reliable scientific knowledge. The success of science in producing new knowledge depended, according to this analysis, on the autonomy of science as a social institution. Merton believed that the autonomy of science was threatened by a widespread anti-science tendency, which for Merton was illustrated by the direct intervention in the affairs of science by totalitarian states. He argued that only in societies that foster semi-independent subgroups, such as the professions, could science flourish. Science required this independence because it depended on the free operation of the scientific ethos. He argued that because this ethos at times conflicts with the institutional goals of the state, it needs to be insulated from direct state intervention. Merton argued that liberal democratic states provided a suitable social environment for science, or as he put it: 'science is afforded opportunity for development in a democratic order which is integrated with the ethos of science' (Merton 1973, p. 269).

Merton's argument was that science functioned according to a set of social norms or 'rules of the game'. These norms function to maximize the contribution of qualified scientists to the production of knowledge, to ensure the reliability of the knowledge generated and to make it as widely available as possible. The 'ethos of science' is the social order that results from the operation of this set of norms. Merton held that the norms were enforced by the social group, and because belonging to the community of scientists is of central importance to a practising scientist (see Gregory, Chapter 1.1 this volume), the norms are a powerful instrument for disciplining individual behaviour:

The ethos of science refers to an emotionally toned complex of rules, prescriptions, mores, beliefs, values, and presuppositions which are held to be binding upon the scientists. Some phases of this complex may be methodologically desirable, but observance of the rules is not dictated solely by methodological considerations. This ethos, as social codes generally, is sustained by the sentiments of those to whom it applies. Transgression is curbed by the internalised prohibitions and by disproving emotional reactions which are mobilized by the supporters of the ethos. Once given an effective ethos of this type, resentment, scorn, and other attitudes of antipathy operate almost automatically to stabilize the existing structure.

Merton (1973, footnote 15, p. 258)

Merton argued that there were four norms that combined to make up the ethos of science: communism, universalism, disinterestedness and organized scepticism; some formulations have five such norms, spelling out the significant acronym CUDOS, which Merton argued was a highly significant motivational factor for scientists. Put simply, *communism* (which later became better known as 'communalism') means that scientists have a duty to share their knowledge freely among the community of scientists. The *universalism* imperative requires scientific statements to be valid for anyone, anywhere. So when judging a statement to be true or false the scientist should neither be inclined to judge statements made by friends and allies to be true, nor more likely to judge statements made by enemies or competitors to be false. In the context of the 1930s, Merton's point was that for the successful functioning of science, any attributions such as nationality or race to the author of a truth-claim should not affect the scientific evaluation of that claim. *Disinterestedness* refers to the proper detachment of the scientist from their work. According to Merton's account of the institutional norms of science, a scientist should not have a vested interest in seeing their particular claim judged to be true.[4] And finally, *organized scepticism* implies that scientists should tend to doubt, and that experimentation and peer review should be carried out with the intention of 'testing' even the most widely held convictions.

While both King, in his capacity as a science policy official, and Merton, the sociologist, emphasize the links between the moral conduct of scientists and public support for science, they understand this relationship in subtly different ways. Merton assumes that science is a unified activity defined by a scientific method that has been stable since its emergence in Europe in the 17th century. For Merton, science is unquestionably to be valued for its capacity to add to the stock of human knowledge, but he does not assume that social support for science is inevitable. In fact, Merton expects there to be some conflict between science and other social institutions, which is why he argues for the autonomy of science. By contrast, King's view is that the public will support science as long as they are made aware of the ethical code followed by scientists. But King also recognizes that scientists may deviate from accepted ethical behaviour, and also that the ethical code extends beyond an expression of the scientific method to include respect for human and animal research subjects, the environmental impact of science and in addition public dialogue.

4. The 'disinterest' norm is somewhat contradicted by Merton's own account of the reward structure in science, according to which professional respect is won by a scientist who is recognized as the first to make a particular discovery. Merton's solution to this apparent contradiction is to make an important distinction between individual motivation and institutional norms, which may at times moderate individual behaviour (Merton 1973, pp. 275–6).

Science communication and ethical norms in science

While both King's prescriptive ethical code and Merton's description of a scientific ethos start from a concern with the place of science in society, both focus on the operation of science as a set of knowledge-making practices. One way to draw out the implications of their approaches to understanding the relationship between science and the wider society is to focus on the theme of communication.

Consideration of science communication in its broadest sense can help in understanding the relationship between the production of scientific knowledge and its public authority. This approach raises two of the most fundamental questions in the study of science in society. How do scientists agree on what should or should not count as scientific knowledge? And, what is the status of science within wider society and how is this status attained and maintained? How scientific knowledge is communicated is key to both these questions. In the first, how scientists successfully communicate their knowledge to other scientists; in the second, the communication between science and its publics.

As this volume and its companion illustrate (Holliman *et al.* 2009) science communication has emerged in recent years as an area of research and practice in its own right. Its origins can be traced to a practical concern with how to disseminate scientific knowledge from specialist communities of scientists to diverse audiences such as policy-makers, journalists and wider publics. However, what started as a set of practical questions about how to communicate science soon bumped up against theoretical questions shared by the field of science and technology studies (STS). Debates about the relationship between what became known as the public understanding of science and public trust in science soon illuminated a much more complex and often contested set of relationships between sciences and publics. For example, whether scientific literacy should be thought of in terms of knowledge of scientific facts or scientific methods; if there is a relationship between public understanding of science and public trust in science; and whether or not public controversy over technologies can be explained in terms of a lack of public knowledge of science (see Wynne 1995; Irwin and Wynne 1996).

Historians of science have argued that since the scientific revolution of the 17th century, the commitment of experimentalists to sharing information openly among their peers has set science apart from other forms of knowledge. Truth claims in science are not supported by experiments alone, but by a process of sharing information about the experimental set-up, the findings and their theoretical interpretation. The operation of science depended on scientists trusting each others' accounts of what they observed. It therefore relied on an implicit ethical code of virtue in public life that pre-dates the development of experimental science (Shapin 1994).

But science did not just depend on existing codes of civility to underwrite the claims made by scientists. One of the key advancements made by scientific practice towards the end of the 17th century was to formalize the mechanisms of scientific communication by which the scientific community as a whole could collectively witness the testing of natural laws. Robert Boyle, one of the foremost experimentalists of the new Royal Society, went to great lengths to demonstrate the workings of his air pump experiment in front of other scholars. He even contributed to developing the literary genre of the scientific paper in order to allow other scholars, who were not physically present, to participate at a distance in the process of witnessing and critically commenting on experimental findings (Shapin 1984).

However, in the context of a renegotiation of the place of science within contemporary society a wider understanding scientific communication than that between fellow scientists is necessary. By analysing the assumptions made about science communication in both King's code and Merton's norms it is possible to differentiate between four modes of science communication. First there is the communication among scientists that is constitutive of the process of knowledge production. Second there is the strategic communication that is part of the process of choosing research directions, whether through choice of career, research topics or approaches to a particular problem. This strategic dimension to science requires communication as scientists align their work in relation to other scientists, to funding opportunities and to potential reward structures. Third, there is the dissemination of science to audiences outside science. And finally there is dialogical communication between scientists and non-scientists about the potential uses and impacts of scientific research (Table 1). I will now discuss each of these four modes of communication as they appear in King's code and Merton's norms.

Table 1 Four modes of science communication

	Content of scientific knowledge	Uses of scientific knowledge
Scientists communicating with scientists	1. Constitutive	2. Strategic
Scientists communicating with non-scientists	3. Disseminating	4. Dialogic

Constitutive communication

Merton's scientific norms are primarily concerned with communication among scientists about the content of scientific knowledge. The importance attached to the open, transparent and honest communication—not just of knowledge claims but the theoretical and material conditions for their production—is illustrated by Merton's account of the scientific ethos. The norm of communism implies a responsibility to share knowledge through communication to other scientists. The norms of universalism and disinterestedness suggest that this communication should be open and free and not favour particular groups of scientists at the expense of others.

The 'Universal Ethical Code for Science' is divided into three sections (see Box 1), which correspond approximately to an ethical underpinning of the scientific method ('rigour'); research ethics concerning the immediate impacts of research on its subjects and the environment ('respect'); and the ethics of relations between science and wider society ('responsibility'). King does not explicitly include an ethic of communication of scientific knowledge among scientists.[5] However, the 'rigour' section of the code concerns a scientist's

5. The omission of a provision concerning communication among scientists might possibly be because it appears obvious to academic scientists whose career depends on publishing the results of their work. Alternatively it could be because King hopes that the code will be taken up by all scientists not just academic scientists and some government and industry scientists are not free to communicate their research due to requirements of commercial or state secrecy.

relations with the wider scientific community and, like Merton's norms, assumes open and transparent communication. According to King's code scientists should help teach one another relevant skills and acknowledge the work of others. The provision that scientists should declare any interests in their research implies that open forms of constitutive communication can function to guarantee the trustworthiness of scientific knowledge.

Strategic communication

The second form of communication, strategic communication, is not explicitly addressed by Merton's four norms. Merton's account asserts that the goal of science is to add to the stock of certified knowledge. He argues that this goal is achieved through the compatibility of the technical means of the scientific method and the social means of the scientific ethos. But this does not address the question of how scientists make choices about what to study. In later work, Merton discusses the reward structure in science. The pursuit of professional recognition is what drives many scientists, according to Merton (1957), and the principal form of recognition is to be credited with being first to make a particular discovery. For Merton, then, strategic communication is the same as constitutive communication. Communicating with fellow scientists in the same discipline allows the scientist to assess the state of the field and make choices about what area of research offers the greatest opportunity to produce new knowledge and therefore gain the recognition of one's peers.

King's ethical code addresses aspects of strategic communication as a distinct form of science communication. The code calls on scientists to 'minimise and justify any adverse effect' of research on people, animals and the environment. This suggests criteria for selection of research methods or perhaps even research topics, keeping in mind how the conduct of scientific research itself might affect others. It also requires scientists to justify any adverse effects, although it is left unclear to whom this justification is made—and importantly, on what terms the attempt to justify a particular course of action might be judged.

Dissemination and dialogue

Consideration of assumptions about communication between scientists and non-scientists can also be used to shed light on the similarity and differences of Merton's account of scientific norms and King's proposed code. The disseminating communication and dialogic communication are the third and fourth modes considered here. Dissemination is the mode that characterizes the post-World War II social contract for science: science communicated *to* the public. Knowledge, technology and the scientifically trained workforce would carry the fruits of scientific research from the origin of their production to wider society. In contrast, dialogue is the mode that characterizes attempts, particularly in the UK since the BSE crisis of the mid-1990s, to restore public trust in science (see Irwin 2009; Stilgoe and Wilsdon 2009).

Merton's mid-20th-century description of the scientific ethos as something particular to science and vulnerable to threats from state intervention leaves no room for discussion of dialogical communication. His argument is that science should be preserved as a semi-autonomous social realm because the social ethos that supports successful scientific practice may at times come into conflict with wider social values.[6] However, Merton does consider the ways in which the wider social context is important for science. The first is that science, if it is to continue as a large-scale activity, requires public support, not least in the form of money. The second is that in order for science to function along the lines described by Merton, the wider social order must be such that it can accommodate the degree of autonomy required by science.

Merton, in his 1938 paper, is cautious about the consequences of disseminating scientific knowledge to publics. He does not assume that wider dissemination would lead to an increase in science's social status. Merton argues that as science becomes increasingly specialized and complex it becomes less intelligible to non-specialists. The dissemination of popularized versions of science that cannot be evaluated by publics encourages a credulity, which opens the door to public acceptance of 'new mysticisms clothed in apparently scientific jargon' (Merton 1973, p. 264). This, suggests Merton, could be one reason for the rise of totalitarian ideologies that in his view undermine the status of science.

For Merton, the main route of disseminating communication is through the application of science in the form of technologies. These may be understood as demonstrating the power of scientific theories in ways that would not be directly ascertainable to publics. However, this is, as Merton points out, a double-edged sword. For when the technologies are considered beneficial then science as a whole may benefit from the reflected glory, but when technologies are seen as detrimental, science as a whole suffers a loss of public support. Here Merton gives the examples of the automation of production processes that disadvantage workers and 'new' military technologies, such as the use of chemical weapons in World War I. Merton criticizes the:

. . . tendency for scientists to assume that the social effects of science *must* be beneficial in the long run. This article of faith performs the function of providing a rationale for scientific research but it is manifestly not a statement of fact. It involves the confusion of truth and social utility which is characteristically found in the nonlogical penumbra of science.

Merton (1973, p. 263)

For King, the wider public are an important audience for the universal ethical code. On the web site of the Department for Business Enterprise and Regulatory Reform, in which the Government Office for Science is located, the code appears in the 'Science and society' section, under the heading of 'Public Engagement and Partnerships'.[7] King has argued that publicizing the code:

. . . provides a simple summary of the values that each of us espouses as practising scientists; it should also demonstrate to the public that scientists take ethical issues seriously.

Our social licence to operate as scientists needs to be founded on a continually renewed relationship of trust between scientists and society. The code has been developed in my Office to help us meet this challenge.

King (2007)

6. Although, as we have already discussed, the motivations of individual scientists are varied, Merton believed that science as a social institution orders individual action.

7. http://www.berr.gov.uk/dius/science/science-and-society/public_engagement/code/index.html.

Conclusion

This chapter set out to explore the relationship between how scientists communicate among themselves, how they communicate with wider publics and the place of science in society. It also describes the recent publication by David King, then the UK government's chief scientific adviser, of a universal ethical code for scientists. The chapter analyses this code, and the assumptions it makes about science communication, by drawing parallels and contrasts with Merton's description of the moral order of science dating back just over half a century.

This chapter has drawn attention to two contrasting codes of scientific practice; while King's code is proscriptive, Merton's is descriptive, though that description relates to the functioning of an idealized scientific community. But the difference is less stark than might seem the case at first. King's code attempts to capture what is already going on, with an optimistic assumption that routine practice is generally ethically sound, because of what he sees as the soundness of the scientific method. But while Merton's work is strongly informed by his historical study of 17th-century England, his two essays (Merton 1938, 1942) are not directly based on empirical work, which may in part explain why the applicability of Merton's norms to modern laboratory life have been so widely questioned (see, for example, Ziman 2000).

A strength of Merton's analysis is that it points to the question of social order; indeed this is a key aspect that has attracted a good deal of insightful writing (see, for example, Barnes 2003). What strikes me as a weakness of Merton's case is the strong assumption that pure science is both a realistic category and necessarily desirable.

King's code has the advantage of explicitly linking scientific work with 'external' considerations of social and environmental uses. His form of communication (as in 'listen to the aspirations and concerns of others') is pleasingly less one-way than Merton's 'dissemination' of scientific knowledge, itself a rather underplayed component. But what to me is the most disappointing weakness in King's argument is the focus on public trust as an end point. As Alan Irwin (2009) argues in the companion volume, the notion of trust being generated as a genial consequence of effective dialogue between science and society is optimistic if not fanciful; effective dialogue will raise fundamental and problematic questions about the direction and governance of science, technology and their institutions.

The case of nanotechnology illustrates this point. In the UK, a range of inventive and ambitious programmes of consultation and dialogue are aiming to ensure an effective dialogue between scientists and different publics.[8] Such discussions included just those problematic areas highlighted by the House of Lords (2000) report as of particular concern—aspects such as health and safety, ethical and societal implications and regulation. Weighty reports (in particular RS/RAE 2004) urge that nanotechnology be the test bed for 'upstream engagement' (see Wilsdon and Willis 2004; Stilgoe and Wilsdon 2009), and conclude by recommending a series of further well-intentioned initiatives involving

---

8. For example, see Stilgoe (2007), Nanojury UK (http://www.nanojury.org.uk/) and Nanodialogue (see for example http://cordis.europa.eu/nanotechnology/).

public and stakeholder dialogue. It is instructive to reflect on King's ethical code in the light of such challenges. The 'responsible communication' that King urges may be laudable but worthy sentiments of that type offer little prospect of an ensuing trust, if (as is the case with nanotechnology) the future direction and focus of research is so uncertain and the individuals and institutions that drive research are (as at present) lacking a mechanism that allows such public dialogue to shape research priorities. Something more urgent and structured than 'seek to discuss' and 'listen to the aspirations of others' may be required. Creating a high expectation of public influence on research priorities that cannot be delivered on is hardly the best precursor for public trust. And as this chapter has argued, trust not only seems an optimistic outcome—it is an unattractively narrow one that fails to do justice to the vision of broader societal influences on contemporary science that many now seek.

Acknowledgement

I would like to acknowledge the support of the Wellcome Trust, which funds my current research on the public dimensions of scientific practice. I would also like to thank Kevin Burchell, Sarah Franklin and Kerry Holden for regular conversations on the topic of public engagement with science.

■ **REFERENCES**

Barnes, B. (2003). Thomas Kuhn and the problem of social order in science. In: *Thomas Kuhn* (ed. T. Nickles), pp. 122–41. Cambridge University Press, Cambridge.

Beck, U. (1992). *Risk Society: Towards a New Modernity*. Sage, London.

Bush, V. (July 1945). *Science: the Endless Frontier*. A report to the President by the Director of the Office of Scientific Research and Development. United States Government Printing Office, Washington, DC.

DIUS (2007). *Rigour, Respect and Responsibility: a Universal Ethical Code for Scientists*. DIUS, London. Available at: **http://www.berr.gov.uk/dius/science/science-and-society/public_engagement/code/index.html**.

Duffy, M. (1986). The Rothschild experience: health science policy and society in Britain. *Science, Technology and Human Values*, **11**, 68–78.

Gibbons, M., Limoges, C., Nowotny, H., Schwartzman, S., Scott, P. and Trow, M. (1994). *The New Production of Knowledge; the Dynamics of Science and Research in Contemporary Societies*. Sage, London.

Highfield, R. and Fleming, N. (12 September 2007). 'Hippocratic oath' for scientists proposed. *The Telegraph: Earth*. Available at: **http://www.telegraph.co.uk/earth/main.jhtml?view=DETAILS&grid=&xml=/earth/2007/09/12/scihipp112.xml**.

Holliman, R., Whitelegg, E., Scanlon, E., Smidt, S. and Thomas, J. (eds) (2009). *Investigating Science Communication in the Information Age: Implications for Public Engagement and Popular Media*. Oxford University Press, Oxford.

House of Lords, Select Committee on Science and Technology (2000). *Science and Society*, Third Report. HMSO, London.

Irwin, A. (2009). Moving forwards or in circles? Science communication and scientific governance in an age of innovation. In: *Investigating Science Communication in the Information Age: Implications for Public Engagement and Popular Media* (ed. R. Holliman, E. Whitelegg, E. Scanlon, S. Smidt and J. Thomas). Oxford University Press, Oxford.

Irwin, A. and Wynne, B. (eds) (1996). *Misunderstanding Science: the Public Reconstruction of Science and Technology*. Cambridge University Press, Cambridge.

Jasanoff, S. (2005). *Designs on Nature: Science and Democracy in Europe and the United States*. Princeton University Press, Princeton, NJ.

Kennedy, D. (2005). Bayh-Dole: almost 25. *Science*, **307**, 1375.

King, D. (20 March 2007). Rigour, respect and responsibility. *The Guardian: Education.* Available at: **http://education.guardian.co.uk/egweekly/story/0,,2037480,00.html**.

McGucken, W. (1984). *Scientists, Society, and the State: the Social Relations of Science Movement in Great Britain, 1931–1947*. Ohio State University Press, Columbus, OH.

Merton, R. (1938). Science and the social order, *Philosophy of Science*, **5**, 321–37.

Merton, R. (1942). Science and technology in a democratic order. *Journal of Legal and Political Science*, **1**, 115–26.

Merton, R. (1957). Priorities in scientific discovery: a chapter in the sociology of science. *American Sociological Review*, **22**, 635–59.

Merton, R. (1973). *The Sociology of Science: Theoretical and Empirical Investigations*. University of Chicago Press, Chicago.

Nowotny, H., Scott, P. and Gibbons, M. (2001). *Re-thinking Science: Knowledge and the Public in an Age of Uncertainty*. Polity Press, Cambridge.

Peters, K. (9 May 2005). Letter from the Council for Science and Technology. Available at: **http://www2.cst.gov.uk/cst/news/**.

RS/RAE (Royal Society/Royal Academy of Engineering) (July 2004). *Nanoscience and Nanotechnologies: Opportunities and Uncertainties*. Royal Society Policy Document 19/04. Royal Society, London.

Shapin, S. (1984). Pump and circumstance: Robert Boyle's literary technology. *Social Studies of Science*, **14**(4), 481–520.

Shapin, S. (1994). *A Social History of Truth: Civility and Science in Seventeenth-century England*. University of Chicago Press, Chicago.

Stilgoe, J. (2007). *Nanodialogues: Experiments in Public Engagement with Science*. Demos, London.

Stilgoe, J. and Wilsdon, J. (2009). The new politics of public engagement with science? In: *Investigating Science Communication in the Information Age: Implications for Public Engagement and Popular Media* (ed. R. Holliman, E. Whitelegg, E. Scanlon, S. Smidt and J. Thomas). Oxford University Press, Oxford.

Wilsdon, J. and Willis, R. (2004). *See Through Science: Why Public Engagement Needs to Move Upstream*. Demos, London.

Wynne, B. (1995). Public understanding of science. In: *Handbook of Science and Technology Studies* (ed. S. Jasanoff, G. Markle, J. Petersen and T. Pinch), pp. 361–88. Sage, London.

Wynne, B. (1996). May the sheep safely graze? A reflexive view of the expert-lay knowledge divide. In: *Risk, Environment and Modernity: Towards a New Ecology* (ed. S. Lash, B. Szerszynski and B. Wynne), pp. 44–83. Sage, London.

Ziman, J. (2000). *Real Science*. Cambridge University Press, Cambridge.

■ FURTHER READING

- Merton, R. (1973). *The Sociology of Science: Theoretical and Empirical Investigations*. University of Chicago Press, Chicago. The following readings all relate to Robert Merton's sociology of science. His major works on science are collected in this book, with very helpful introductions to each section by Norman Storer.

- Turner, S. (2007). The social study of science before Kuhn. In: *The Handbook of Science and Technology Studies* (ed. E. Hackett, O. Amsterdamska, M. Lynch and J. Wajcman), pp. 33–62. MIT Press, Cambridge MA. For an introduction to the wider context for Merton's work this recent chapter is very helpful, although Merton is only one of many theorists that are discussed.

- Mulkay, M. (1976). Norms and ideology in science. *Social Science Information*, **15**, 637–56. The most widely read of the critiques of Merton's account of the normative structure of science is this 1976 paper by Mulkay. Here he criticizes Merton for basing his account of how science works on the writings and reflections of great scientists, rather than a more detailed study of the day-to-day world of science. Mulkay argues that the norms Merton identifies are self-serving myths that leading scientists perpetuate in order to maintain independence from external scrutiny.

- Cole, S. (2004). Merton's contribution to the sociology of science. *Social Studies of Science*, **12**, 829–44. This synoptic account of the legacy of Merton's sociology of science, as well as some of its shortcomings, is written by a former student of and one time collaborator with Merton.

- Barnes, B. (2003). Thomas Kuhn and the problem of social order in science. In: *Thomas Kuhn* (ed. T. Nickles), pp. 122–41. Cambridge University Press, Cambridge. In recent years some authors have been returning to the questions about the social order of science posed by Merton. This is a good example of such work.

■ USEFUL WEB SITES

- **Nanoscience and Nanotechnologies: opportunities and uncertainties**: http://www.nanotec.org.uk/. This is the web site for the study on nanotechnology carried out by the Royal Society and Royal Academy of Engineering (2004). The UK government commissioned the study in 2003 to assess the current state of nanoscience and nanotechnology research and assess the policy implications posed by likely technological developments.

- **Council for Science and Technology**: http://www2.cst.gov.uk/cst/business/nanoreview.shtml. The Council for Science and Technology (CST) reviewed the government's progress in meeting the commitments it made in response to the RS/RAE (2004) study. You can download a copy of the CST's report from this web site.

- **Rigour, respect and responsibility: a universal ethical code for scientists**: http://www.berr.gov.uk/dius/science/science-and-society/public_engagement/code/index.html. David King's ethical code for scientists can be found at this UK government web site. There is some background material on the 'Universal ethical code', and links to the Council for Science and Technology (CST) 2005 consultation on how the code could be used. You can also download a copy of the leaflet referred to in the chapter. (It is worth noting that the code is found under the 'Public Engagement and Partnerships' heading of the 'Science and Society' section of the web site. This reinforces the argument of the chapter that the code is primarily aimed at public communication rather that changing scientific practice.)

• **History of the US National Science Foundation (NSF): http://www.nsf.gov/about/history/index.jsp**. This section of the NSF web site was set up to celebrate the 50th anniversary of the foundation of the NSF in 1950. If you click through to the 'Publications' page you will find several interesting documents that shed light on the post-World War II 'social contract' for science in the USA. These include an online version of Vannevar Bush's *Science: the Endless Frontier* (which is discussed in the introduction to this chapter), and also a brief history of the NSF that talks about the transition from wartime research to the establishment of the NSF at arm's length from government in order to ensure the freedom of scientists to set their own research agendas.

1.3

Patents and the dissemination of scientific knowledge

Charlotte Schulze

This chapter introduces patents in the context of science communication. The purpose, format and content of patent documents are compared with those of peer-reviewed scientific articles covering the same scientific work. The use of legal terminology has largely been avoided, although a few specialist terms have had to remain. These might appear arcane, but they are illustrative of the specialist jargon and their meaning should become clear in the context in which they are introduced.

The patenting of the results of scientific research, including those carried out at publicly funded research institutes, has significantly increased in recent years. At the same time the world-wide patent literature has become much more readily accessible to all. These developments are making it essential for many research scientists to develop patent literacy skills, as they need to be aware of patenting activities in their own field of research.

What is intellectual property?

Intellectual property (IP) is about the idea of making knowledge and information 'ownable' (May and Sell 2006, p. 5). According to the World Intellectual Property Organization (WIPO):

Intellectual property refers to creations of the mind: inventions, literary and artistic works, and symbols, names, images, and designs used in commerce.

WIPO (2007)

The UK Intellectual Property Office has provided a slightly more detailed, but still very much simplified definition (Box 1). It lists copyright, designs, patents and trade marks as the four main types of IP, but also refers to a range of other, more specialized, types such as trade secrets and geographical indications.

BOX 1 WHAT IS INTELLECTUAL PROPERTY?

Intellectual property (IP) can allow you to own things you create in a similar way to owning physical property. You can control the use of your IP, and use it to gain reward. This encourages further innovation and creativity.

The four main types of IP are:

- **Copyright**: copyright protects material, such as literature, art, music, sound recordings, films and broadcasts.
- **Designs**: designs protect the visual appearance or eye appeal of products.
- **Patents**: patents protect the technical and functional aspects of products and processes.
- **Trade marks**: trade marks protect signs that can distinguish the goods and services of one trader from those of another.

However, IP also covers trade secrets, plant varieties, geographical indications, performer's rights and so on. Often, more than one type of IP may apply to the same creation.

UK Intellectual Property Office (http://www.ipo.gov.uk/whatis.htm)

Of greatest relevance for the natural sciences in terms of IP protection are *patents*. Box 2 explains, in 'non-legal' terms, what patents protect and which rights they confer. An invention needs to have certain properties in order to be eligible for protection as a patent: it needs to be novel, have an inventive step, exhibit industrial applicability and must *not* fall into any of the 'exempted subject matter' categories. For example, 'naturally occurring' animal or plant varieties cannot be patented. However, as we shall see, genetically modified plants or animals can be patented.

BOX 2 WHAT IS A PATENT?

A patent protects new inventions and covers how things work, what they do, how they do it, what they are made of and how they are made. It gives the owner the right to prevent others from making, using, importing or selling the invention without permission.

Your invention must:

- be **new**
- have an **inventive step** that is not obvious to someone with knowledge and experience in the subject
- be capable of being **made** or **used** in some kind of industry [its 'industrial applicability']
- **not** be: a scientific or mathematical discovery, theory or method; a literary, dramatic, musical or artistic work; a way of performing a mental act, playing a game or doing business; the presentation of information, or some computer programs; an animal or plant variety; a method of medical treatment or diagnosis; against public policy or morality.

UK Intellectual Property Office (http://www.ipo.gov.uk/whatis/whatis-patent.htm)

In fact, patent law is extremely complex and its interpretation requires specialist expertise. For example, whilst patent law in the UK is codified in the Patents Act 1977, it is interpreted and evolving through case law, patent office practice, European legislation and other such influences. For the purposes of this chapter it suffices to give a broad and very simplified overview. It is important to keep in mind that although most countries have broadly comparable intellectual property regimes they are not identical, and important—if sometimes minor—differences between them do exist. Therefore, over-generalizations have to be made with caution.

The term 'intellectual property' as a unifying concept only came into general use in the second half of the 20th century (May and Sell 2006, p. 18). Prior to this reference was made to 'industrial property' or to the individual rights such as trade marks or copyright. The Agreement on Trade-Related Aspects of Intellectual Property Rights (TRIPS) of 1994 represents:

... the first truly global intellectual property agreement (with robust enforcement mechanisms through the World Trade Organization) and the first to cover all forms of intellectual property rights under the same set of legal mechanism

May and Sell (2006, p. 19)

In the words of the World Trade Organization (WTO):

... the WTO's TRIPS Agreement is an attempt to narrow the gaps in the way these rights are protected around the world, and to bring them under common international rules. It establishes minimum levels of protection that each government has to give to the intellectual property of fellow WTO members

WTO (2007)

Thus, intellectual property rights exist on a national (or regional) level and the ratification of international agreements such as TRIPS attempts to unify and strengthen legal protection world-wide. In this way new inventions can be patented world-wide so others cannot 'work the invention' (i.e. do what the patent prohibits them from doing) without permission by the patent owner for the duration of the patent—usually 20 years.

The following case study illustrates the story of an invention for which near world-wide patent protection is being sought.

Case study: drought resistant, genetically modified plants

On 26 November 2007 *Nature News Online* reported that plant biologist Professor Eduardo Blumwald of the University of California Davis and co-workers had produced a genetically modified (GM) tobacco plant that could survive prolonged dry periods and, in addition, thrive on 70% less water than ordinary tobacco plants (Ledford 2007). This report announced the publication of the corresponding scientific research paper that came a few days later (Rivero *et al.* 2007).

A press release by the University of California on the same day revealed that a patent application on this technology had been filed and was pending 'in the US and a number of foreign countries' (University of California Newsroom 2007). An exclusive arrangement had been made over the patent rights between the university and Arcadia Biosciences, an agricultural biotechnology company, who had provided financial support for the research (University of California Newsroom 2007).

A patent search reveals that Professor Blumwald was already a named inventor on a number of recent patents and patent applications. One of them is entitled 'Drought-resistant plants' (Gepstein *et al.* 2006; patent number WO/2006/102559), which carries the international publication date 28 September 2006 and an international priority date of 21 March 2005.[1]

Due to the potential significance and newsworthiness of this new technology, as well as the promotional strategies employed by the university (see Allan 2009; Trench 2009 for discussion), the story was reported by science journalists world-wide. Arcadia Biosciences then published a press release on 10 December 2007 (Arcadia Biosciences 2007) reporting the successful completion of some field trials involving the transgenic, drought-resistant plants. This press release was, in turn, covered by a number of organizations reporting to financial markets.

Thus we can see how a promising piece of scientific innovation was reported as a news item by one of the most prestigious science journals, *Nature* (Ledford 2007), as a press release and as a peer-reviewed scientific article (Rivero *et al.* 2007). However, only through the university's press release do we learn about the existence of a patent application covering the invention and the fact that Arcadia Biosciences owns the exclusive rights over it. This patent was filed approximately 21 months before the scientific paper, press release or news article appeared. In effect, it was published—and therefore became generally accessible—some 14 months prior to the scientific paper.

Comparing forms of science communication

Let's now compare the content of the paper and patent, respectively (Table 1). The paper by Rivero and co-workers appeared in the *Proceedings of the National Academy of Sciences* (*PNAS*), a prestigious, peer-reviewed weekly journal. It follows a standard format for scientific papers and is six pages long. The paper explains that many plants react to periods of drought by wilting and shedding of leaves, a process which can rapidly lead to plant death. The authors speculated that prolonged drought switches on an 'inappropriate' process of programmed cell death which causes the wilting and shedding of leaves. They hypothesized that if they were able to delay this 'drought-induced senescence of leaves' by genetically modifying plants, it may be possible to enhance the plant's drought tolerance.

Through careful analysis of the literature in the field, Rivero *et al.* (2007) identified a signalling pathway that plays a role during the very early stage of senescence. They chose tobacco plants and inserted a specific gene driven by a particular promoter with the

---

1. Each patent has a 'priority date' which is the date from which protection is conferred and for which the essential criteria (novelty, inventive step etc, see Box 2) are assessed. Patent applications are only published ('publication date') after a period of several months (depending on the corresponding legislation).

Table 1 Comparing the scientific paper with the patent application

	Scientific paper	Patent application[a]
Title	Delayed leaf senescence induces extreme drought tolerance in a flowering plant	Drought-resistant plants
Authors/inventors	R.M. Rivero, M. Kojima, A. Gepstein, H. Sakakibara, R. Mittler, S. Gepstein, E. Blumwald	S. Gepstein, A. Gepstein, E. Blumwald
Publication date	04.12.2007 (online: 28.11.2007)	28.09.2006 (international publication date) 21.03.2005 (priority date)
Structure	**Title** (and authors, affiliations, contact details, declaration of no conflict of interest) **Summary** (and keywords) **Introduction** **Results** (including 6 figures) **Discussion** **Methods** (and acknowledgement of sponsors) **References** **Supporting information** (a further 7 figures are referred to in the text; they are available online free of charge)	**Bibliographic data** (international publication date, intern. publication number, intern. patent classification, intern. application number, intern. filing date, filing language, publication language, priority data, applicant, inventors, agents, designated states (national and regional protection)) **Title** **Abstract** (2 sentences) **Description** (comprising sections entitled 'Cross-references to related applications', 'Background of the invention', 'Brief summary of the invention', 'Brief description of the drawings', 'Detailed description of the invention' (including definitions, detailed methodology and results relating to specific embodiments and examples) **Claims** (includes 10 claims) **Drawings** (includes 4 figures)

[a] This refers to the facsimile of the published patent application with international publication number WO/2006/102559 A2 (Gepstein *et al.* 2006).

aim to produce transgenic plants ($P_{SARC}::IPT$-expressing plants) with the desired effect of delaying onset of senescence. Of the plants they produced in this way, they selected those that showed increased drought tolerance. This is how they summarize their findings:

. . . our results show that the suppression of drought-induced leaf senescence resulted in outstanding drought tolerance. The plants had vigorous growth after a long drought period that killed the control plants. These plants had minimal yield loss when watered with only 30% of the amount of water used under control conditions. These results are exciting because they indicate that, in addition to increased drought tolerance, the expression of $P_{SARC}::IPT$ in plants, with the concomitant increase in water use efficiency, could facilitate the development of transgenic crops able to grow with reduced irrigation without significant yield penalties, contributing to significant savings in irrigation water

Rivero *et al.* (2007, p. 19635)

The article is characterized by the fact that the reasoning behind the experimental design and the chosen approach are explained in some detail by frequently referring to the scientific literature in the field. The authors describe how they have produced the transgenic tobacco plants and how the properties of these plants have changed as a result. They then describe an array of additional experiments that serve to demonstrate conclusively that their chosen methodology, their observations and their interpretation of the results are reliable and sound and that alternative explanations of the observed phenomena can be ruled out.

The corresponding patent (Gepstein *et al.* 2006) has quite a different structure and content (Table 1). In contrast to the scientific paper its length is not determined by journal constraints but by the subject matter covered. The brief section on background information refers to some scientific literature as well as to some patents. These are mostly different from those described in the paper and they serve to describe and identify the 'prior art' relating to the invention.

The detailed description of the invention begins with over three pages of definitions in order to leave no doubt about what exactly is claimed. For example, the term plant:

[. . .] includes whole plants, shoot vegetative organs/structures (e.g. leaves, stems and tubers), roots, flowers, and floral organs/structures . . . seed . . . and fruit . . . plant tissue . . . and cells . . . and progeny of the same. The class of plants that can be used in the method of the invention is generally as broad as the class of higher and lower plants amenable to transformation techniques, including angiosperms . . . gymnosperms, ferns, and multicellular algae. It includes plants of a variety of ploidy levels. . . .

Gepstein et al. (2006, p. 4)

The purpose here is to ensure that anything from which a plant may be generated as well as any plant species 'amenable to transformation techniques' are included in the definition and do not fall outside the patent's scope. Some 15 relevant terms are defined in a similarly comprehensive way.

This is followed by 10 pages of description of methods and results, including several pages of DNA and protein sequences. In contrast to the *PNAS* paper, no 'control experiments' are included, as the purpose here is not to convince a sceptical scientific audience of the merits of this study but rather to carefully and exhaustively delineate and define for what exactly protection is being sought. Nobody should be able to 'work the invention' without breaching the patent once it is granted and in force. Great care is taken to include, for example, alternative means to achieve the same effect: transgenic plants containing specific genes and promoters can be produced in a variety of different ways and all of these need to be covered by the patent (Gepstein 2006, pp. 7–9).

Frequent reference is made to specific 'embodiments' and 'examples' in order to demonstrate that it is not an abstract idea that is being claimed here, but that the invention has actually been put into practice, which is an essential requirement for patent protection (see 'industrial applicability' in Box 2). Although the patent explicitly refers to all plants as shown above, the 'embodiments' only refer to tobacco plants—as in the *PNAS* paper. Although other species are mentioned as well, barely any experimental data from any other species have been included.

The claims, shown in Box 3, define the scope for which protection is sought. Whereas the other parts of the patent are drafted in a style that is similar to that of scientific papers and can be understood by those familiar with the research area, the information contained in the claims is more difficult to interpret. These claims need to fulfil a specific legal purpose and are likely to have been drawn up with the help of a patent agent rather than by the scientist inventors (Crespi 2004). They are to be read in conjunction with the information provided in the description.

> **BOX 3** CLAIMS OF PATENT APPLICATION WO/2006/102559 A2[2]
>
> 1. A method of preparing a plant resistant to drought stress, the method comprising: introducing into a population of plants a recombinant expression cassette comprising a SARK promoter operably linked to a nucleic acid sequence encoding a protein involved in cytokinin synthesis; and selecting a plant that is resistant to drought stress.
> 2. The method of claim 1, wherein the step of introducing is carried out by a sexual cross.
> 3. The method of claim 1, wherein the step of introducing is carried out using *Agrobacterium*.
> 4. The method of claim 1, wherein the SARK promoter is from *Phaseolus vulgaris*.
> 5. The method of claim 1, wherein the SARK promoter is at least 95% identical to SEQ ID NO:1.
> 6. The method of claim 1, wherein the protein involved in cytokinin synthesis is isopentenyl transferase.
> 7. The method of claim 6, wherein the nucleic acid sequence encoding the isopentenyl transferase is from *Agrobacterium*.
> 8. The method of claim 7, wherein the isopentenyl transferase is at least 95% identical to SEQ ID NO:3.
> 9. The method of claim 1, wherein the plant is a dicot.
> 10. The method of claim 9, wherein the plant is tobacco.

In summary, there are similarities and differences in the content and purpose of the two parallel publications (Crespi 1998, 2004). They have two important aspects in common. Firstly, they serve to claim priority, albeit of a different type, over a piece of scientific innovation. For a patent to fulfil the requirement of 'novelty' (see Box 2), none of its components is allowed to be in the public domain prior to its priority date. This is why the authors made sure not to publish any of their findings prior to having obtained sufficient data to file a patent application and secure a priority date. Similarly, one of the ultimate aims for academic researchers is to be the first to publish scientific discoveries in highly respected, peer-reviewed journals. This serves to claim priority and therefore respect in front of their peers. The earlier published patent clearly does not fulfil this purpose.

Secondly, both publications also have in common that they include an 'enabling disclosure', i.e. a detailed description of the invention that allows a specialist in the field to repeat the experiments. If someone can successfully demonstrate that a patent does not include sufficient detail to allow a 'person skilled in the art' to work the invention, it can be found to be invalid. Equally, if fellow scientists are unable to reproduce the findings published in a scientific paper, this could result in the need to publish a 'correction' or the journal may even withdraw it ('retraction'). This would be damaging for the reputation of the scientists involved. However, in each case, as we have seen, the nature of the disclosure is slightly different.

In contrast to these similarities, the article and patent differ quite significantly in content and purpose. The former needs to persuade a sceptical scientific community, whereas

---

2. This refers to the facsimile of the published patent application with international publication number WO/2006/102559 A2. (Gepstein *et al.* 2006, p. 17).

the latter needs to effectively demarcate intellectual property rights. Their respective publication dates are also quite different. In this case, the science news story only breaks in conjunction with the publication of a scientific paper, but did not do so after the earlier publication of the patent application, which essentially contained all the same information. There are, however, also specialist journals who report on innovative patents. *Nature Biotechnology* is an example, reporting to a readership of scientists, the biotech industry and possibly financial markets. Thus, both types of publications are disseminated to different audiences and for different reasons.

Scientists' use of patent searches

The dual systems of patenting and scientific journal publication exist in parallel, although most academic research scientists pay little attention to patents (Seeber 2007). For example, it is extremely rare to find scientific papers that cite patents, whereas the reverse is the rule (Glänzel and Meyer 2003, quoted in Seeber 2007, p. 2418). One of the reasons has been that patents are essentially legal documents and as such are somewhat difficult to read for non-specialists. Scientists routinely undertake literature searches in their field of interest to keep abreast of developments. However, gleaning information from patents requires the use of quite separate data bases, possibly even the visit to a specialist library or instruction of a professional searcher. For example, neither of the two most popular free biomedical data base portals, PubMed and Google Scholar, currently (2008) reference patent data (Seeber 2007, p. 2418).

This situation is now changing. Patent searching from the desktop has recently become much easier. Seeber (2007) describes how the free internet-accessible scientific search engine, Scirus, can access more than 21 million patents. A search can easily be combined with other literature data base portals (for example, PubMed), making routine daily or weekly searches feasible, free of charge and without too much effort. The major national and regional patent offices, such as the US Patent and Trademark Office (USPTO) or the European Patent Office (EPO), also offer free access to extensive online data bases with search and even translation facilities. These facilities are constantly being improved and upgraded and can be expected to become more versatile in the future.

What reasons might scientists have to read patents more frequently? This will largely depend on the area of research and specific relevance. Seeber (2007, p. 2420) mentions the following three reasons why life scientists might read patents:

1. Some information appears earlier in patents than in scientific journals (as we saw in the case study).
2. Patents may contain valuable data that never appear in the literature. The inventor might not have enough data to publish a full paper or may simply not be interested in it.
3. Patents are a source of 'hard-to-get information from commercial suppliers' (for example about the exact composition of laboratory reagents where the supplier does not reveal this).

There is a further reason: universities and other publicly funded research institutions are increasingly taking out patents, meaning that scientists need to become more aware of them as part of their routine research practices.

Recent trends: the increase of patenting activities by universities

Scientific innovation conducted by industry is inherently associated with closely guarded intellectual property rights with the aim of recovering research and development expenses and generating a profit for the company in question. However, as the case study exemplifies, patents held by publicly funded institutions such as universities are now a routine occurrence, particularly in certain scientific disciplines, for example biotechnology, and in certain countries. For example, the passing of the 1980 Bayh–Dole Act, which enabled universities to take title to inventions developed with federal funding, was viewed as an important trigger for this increased commercialization in the USA (Verspargen 2006, p. 612). However, recent research has identified a number of other important factors, such as the increase in patentable subject matter and the rapid technological advances in some sectors, for example biotechnology (Nelson 2001; Mowery *et al.* 2001; Verspargen 2006, p. 622). On average, patents in biotechnology and the health sector represent the largest single group of patents within university portfolios of OECD countries (OECD 2003, pp. 9–19). In the USA at least, life science patents also generate the biggest share of revenue for universities (Agres 2003, pp. 55–6).

In European countries the situation is somewhat different (Verspargen 2006). In the UK, as in the USA, the government is strongly committed to an extensive programme of knowledge transfer from the science base at universities to industry (DTI, 2001).[3] Recent surveys on university commercialization activities and university–industry interactions in the UK demonstrate significant levels of activity with respect to invention disclosures, patent applications, issued patents, licensing activities and the formation of spin-off companies (UNICO 2005). In general, UK universities compare favourably with those in the USA except that less income is generated through licensing.

Other European countries are following similar avenues. In Germany, for example, the so-called 'professor's privilege'[4] was abolished in 2001 and universities are now actively pursuing technology transfer (Verspargen 2006, p. 619). In contrast, in other countries, e.g. Finland, Iceland and Sweden, the 'professor's privilege' still exists or has been newly implemented (e.g. Italy) (Verspargen 2006, p. 619). Comparative research addressing potential differences between such countries with respect to technology transfer from universities is scarce, largely due to the lack of empirical data (Verspargen 2006; Geuna and Nesta 2006; Geuna and Mowery 2007). There has also been very little work on

3. This department has recently undergone reorganization and is currently (2008) known as the Department for Business, Enterprise and Regulatory Reform (BERR).

4. The right of a university academic to take title to inventions made during employment.

whether Bayh–Dole-like legislation has actually led to more efficient knowledge transfer from universities to industry (Verspargen 2006).

Nonetheless, many European countries, Australia and Japan have followed the US example in recent years in introducing Bayh–Dole-like legislation (Monotti and Ricketson 2003; Kneller 2007) and this subject matter has triggered much controversy. The debate began in the USA and has recently spread to Europe. It is characterized by the diversity of stakeholders involved and the range of approaches and vested interests they bring into it. Policy-makers, legal scholars, economists, other social scientists, business analysts, research funding bodies, ethicists and scientists themselves are just some examples of those involved. Here I will briefly introduce a few discussion points that have been raised and that may be of interest in the context of science communication. These examples are certainly not representative of the debate as a whole, but they should serve to illustrate some of the issues and viewpoints put forward and give an impression of the complexity of this debate. It is important to stress that university technology transfer activities affect different scientific disciplines to quite different degrees and therefore over-generalizations should be avoided.

For research scientists who are affected by these changes the new policies have meant that they are obliged to disclose any discoveries they have made to their university employer prior to communicating them at conferences or in the shape of a scientific paper if they have the potential to be protected as intellectual property. This may represent little change from the situation that existed before, as scientists tend to be protective of their ongoing research until they are about to publish it. However, it may also mean that they have to delay publication in order to seek patent protection and that frank exchanges with colleagues can only take place under the guard of confidentiality agreements. In a highly competitive research environment, delaying publication could mean that a competing group publishes first and as a result one's own prospects to be associated with the discovery through publication in a highly respected, peer-reviewed journal are severely diminished or even rendered obsolete. This can have a serious impact on a scientist's career, possibly disproportionately so for more junior scientists who need to demonstrate productivity to further their career prospects ('publish or perish').

An obligation to keep information *secret* is in direct contrast to the notion of *openness* and the obligation to disseminate knowledge obtained through public funding. Shorett *et al.* (2003) writing about the changing norms of the life sciences, have put it like this:

. . . concerns about eventual product development resulting from scientific work are increasing in importance, whereas open publication of data is not.

Shorett *et al.* (2003, p. 123)

In this respect, it has been argued that Merton's CUDOS norms of science might be at risk, if commercial interests become increasingly important (see Doubleday, Chapter 1.2 this volume; Ziman 1996, for discussion).

Another controversial issue concerns a particular feature codified in many patent regimes: the so-called 'research exemption'—'Academic researchers have regularly ignored patents on key technologies' in order to pursue their research objectives (Yancey and Stewart 2007, p. 1225). In fact, in some disciplines it has become nearly impossible not to infringe patents in the routine pursuit of research. In biotechnology, for example, 'infringement happens routinely at every university' (Yancey and Stewart 2007, p. 1226). The 'research exemption' allows the infringement of a patent under some well-defined conditions—for example to prepare the production of a 'generic' drug where a patent is about to expire:[5]

Until recently, many university researchers incorrectly believed that the research exemption . . . extended to fundamental research in general and that they could freely use IP in their research without regard to infringement.

Boettiger and Bennett (2006, p. 321)

Recent US case law, however, has narrowed the research exemption 'beyond any practical use', as has been argued by Boettiger and Bennett (2006). Furthermore, the tendency of universities to produce transferable technology and reap income from licence agreements and start-up successes, makes it 'arguably less defensible for . . . researchers to infringe with impunity' (Yancey and Stewart 2007, p. 1227). As a consequence, *liability for infringement* is a real threat, even for university scientists who pursue non-transferable research but who are using technologies that are patented.

Connected with the issue of the research exemption is the potential that increased patenting activity in general could lead to a blockade of future research (Verspargen 2006, p. 617):

This is especially relevant in areas where progress is to a large extent cumulative, i.e. where new results build upon old research in a strong way. A contemporary case is in the life sciences, where, for example, tools for genetic sequencing are of utmost importance for any research being undertaken in the area. Where these tools are patented, access to them will be restricted, and this will have a negative impact on scientific progress.

Verspargen (2006, p. 617)

An example of such a situation is represented by the 'Golden Rice' project. The aim behind creating this 'humanitarian crop' was to produce pro-vitamin A-enriched rice that would combat blindness and malnutrition in developing nations (Yancey and Stewart 2007, p. 1226). To engage in this project required access to over 40 US patented technologies and would not have been economically viable had the companies not waived their licence fees. Clearly this is an issue which reaches beyond the scope of university research. However, since many university discoveries emerge from 'basic' research:

Such 'basic' discoveries are more likely to have an impact on a whole range of subsequent 'applied' research topics, and thus have a strong potential for blocking such future research.

Verspargen (2006, p. 617)

Increased university technology transfer may also lead to a change in the strategic behaviour of universities:

When patents are an increasingly important output for universities, there may be an incentive to do research in those areas where patents are easily obtained.

Verspargen (2006, p. 617)

---

5. A generic equivalent of a patented drug is one that achieves the same effect, but costs less to make, partly because it has not incurred research and development costs. It is only allowed to come on market when the patent has expired. However, the 'research exemption' may allow all preparations for this moment to take place prior to expiry of the patent.

Although some scholars have suggested that such a strategic shift can be observed in US universities as a consequence of the Bayh–Dole Act, others have disputed this and further research is needed to allow any conclusions to be drawn (discussed in Verspargen 2006, pp. 622–4).

Ethical dimensions

A number of the aspects listed so far have strong ethical dimensions as well. It would, for example, probably be considered unethical if the actions of a publicly funded research institution led to a blockade of future research or if secrecy prevented the dissemination of knowledge generated through public funds.

There are a host of additional issues that have emerged primarily as ethical concerns. Some of these can be categorized as issues to do with 'patentable subject matter': objections are being raised against the possibility to patent certain 'inventions'. Box 2 illustrates that there are a number of things that cannot be patented according to UK law. However, scientific innovation progresses fast and it usually takes years for the interpretation of patent law (for example, through case law) to adapt to these changes. In the meantime many inventors will file patent applications for such novel inventions, where it is uncertain whether or not they may eventually fall into any of the exempted categories.

The patenting of human stem cells is an example here. A brief commentary in *Nature* illustrates how and why a scientist 'decided that a legal fight would advance science' (Loring 2007). Working on the development of human stem cell lines, Loring became aware that she was at risk of infringing a very broad patent that had been granted for primate embryonic stem cells and that the patent owner had the intention to exploit the monopoly and charge all researchers for licences. The patent holder is a university-associated organization and 'requires a license from every academic investigator who uses human embryonic stem cells' (Loring 2007). Loring is currently (2008) fighting to have this and a few other patents re-examined—and eventually invalidated—by the USPTO as she believes that they should not have been granted in the first place as they are extremely broad. She feels obliged 'to make sure that our research can benefit the society that supports it' (Loring 2007).

This raises another ethical issue, namely whether the monopolization of the fruits of publicly funded research is eventually benefiting those who funded it. Or, at a more global level, are drugs or improved crops developed in this way eventually becoming available to those in need?

Clearly, these are issues which go well beyond what can be dealt with in this chapter. However, scientists who find themselves in the middle of such dilemmas are voicing their opinions and trying to find solutions. Yancey and Stewart conclude as follows:

The original proponents of patent protection could not have foreseen a world in which the very building blocks of life could be patented or farmers could be prevented from saving seeds from year to year[6], but our courts, regulators and political leaders are certainly aware of it. . . .

6. Because they are growing genetically modified crops and the owner of the patent rights requires purchase of seeds.

It may prove that no silver bullet exists, but with open source solutions, pressure from open science advocates . . . and open licensing from universities, anticommons effects can hopefully be avoided or minimized.

Yancey and Stewart (2007, p. 1228)

Conclusion

This chapter has highlighted the different roles that patents and scientific articles play in science communication. Both formats serve specific purposes and the dissemination of information through them is carefully and strategically managed in each case. As we have seen, the work of public research scientists is increasingly affected by patenting issues—hence there is a need for many scientists to acquire patent literacy skills.

Patents are a source of innovative scientific knowledge, and this information is now becoming much more readily accessible. However, there is the potential for patents to prevent others from applying that knowledge. This trade-off is a major element of any patent system: it aims to provide incentives for scientific research and innovation through ownership of intellectual property, while guaranteeing the flow of knowledge through full disclosure and limiting the duration of ownership rights (Verspargen 2006, p. 611). Striking that balance is a challenge and scientists may well find themselves confronted with new ethical dilemmas.

■ REFERENCES

Agres, T. (2003). The fruits of university research. *The Scientist*, **17**, 55–6.

Allan, S. (2009). Making science newsworthy: exploring the conventions of science journalism. In: *Investigating Science Communication in the Information Age: Implications for Public Engagement and Popular Media* (ed. R. Holliman, E. Whitelegg, E. Scanlon, S. Smidt and J. Thomas). Oxford University Press, Oxford.

Arcadia Biosciences (2007). *Arcadia Biosciences Reports Progress in Drought-tolerance Crop Development*. Press release, 10 December. Available at: **http://www.arcadiabio.com/media/pr/0021.pdf**.

Boettiger, S. and Bennett, A.B. (2006). Bayh-Dole: if we knew then what we know now. *Nature Biotechnology*, **24**, 320–3.

Crespi, R.S. (1998). Patenting for the research scientist. *Trends in Biotechnology*, **16**, 450–5.

Crespi, R.S. (2004). Patenting for the research scientist: an update. *Trends in Biotechnology*, **22**, 638–42.

DTI (Department of Trade and Industry) (2001). *White Paper: Excellence and Opportunity – a Science and Innovation Policy for the 21st Century*. HMSO, London.

Gepstein, S., Gepstein, A. and Blumwald, E., University of California (US). (2006). *Drought-resistant Plants*. Patent WO/2006/102559, 28 September.

Geuna, A. and Mowery, D. (2007). Publishing and patenting in US and European universities. *Economics of Innovation and New Technology*, **16**, 67–70.

Geuna, A. and Nesta, L.J.J. (2006). University patenting and its effect on academic research: the emerging European evidence. *Research Policy*, **35**, 790–807.

Glänzel, W. and Meyer, M. (2003). Patents cited in the scientific literature: an exploratory study of 'reverse' citation relations. *Scientometrics*, **58**, 415–28.

Kneller, R. (2007). The beginning of university entrepreneurship in Japan: TLOs and bioventures lead the way. *Journal of Technology Transfer*, **32**, 435–56.

Ledford, H. (2007). Researchers engineer drought-resistant plants. Genetically modified tobacco doesn't bite the dust. *Nature News* **http://www.nature.com/news/2007/071126/full/ news.2007.289.html**.

Loring, J. (2007). Stem cells. Commentary: a patent challenge for human embryonic stem cells. *Nature Reports. Stem Cells* **http://www.nature.com/stemcells/2007/0711/071108/full/ stemcells.2007.113.html**.

May, C. and Sell, S. (2006). *Intellectual Property Rights: a Critical History*. Lynne Rienner Publishers, Boulder, CO.

Monotti, A.L. and Ricketson, S. (2003). *Universities and Intellectual Property. Ownership and Exploitation*. Oxford University Press, Oxford.

Mowery, D.C., Nelson, R.R., Sampat, B.N. and Ziedonis, A.A. (2001). The growth of patenting and licensing by U.S. universities: an assessment of the effects of the Bayh-Dole Act of 1980. *Research Policy*, **30**, 99–119.

Nelson, R. (2001). Observations of the post-Bayh-Dole rise of patenting at American universities. *Journal of Technology Transfer*, **26**(1–2), 13–19.

OECD (2003). *Turning Science into Business. Patenting and Licensing at Public Research Organisations*. OECD, Paris.

Rivero, R.M., Kojima, M., Gepstein, A., Sakakibara, H., Mitter, R., Gepstein, S. and Blumwald, E. (2007). Delayed leaf senescence induces extreme drought tolerance in a flowering plant. *Proceedings of the National Academy of Sciences of the USA*, **104**, 19631–19636.

Seeber, F. (2007). Patent searches as a complement to literature searches in the life sciences – a 'how to' tutorial. *Nature Protocols*, **2**, 2418–28.

Shorett, P., Rabinow, P. and Billings, P.R. (2003). The changing norms of the life sciences. *Nature Biotechnology*, **21**, 123–5.

Trench, B. (2009). Science reporting in the electronic embrace of the internet. In: *Investigating Science Communication in the Information Age: Implications for Public Engagement and Popular Media* (ed. R. Holliman, E. Whitelegg, E. Scanlon, S. Smidt and J. Thomas). Oxford University Press, Oxford.

UNICO (University Companies Association) (2005). *UK University Commercialisation Survey: Financial Year 2004*. Available from the University Companies Association on request (**http://www.unico.org.uk/contact.htm**).

University of California Newsroom (2007). *New Drought-tolerant Plants Offer Hope Worldwide*. Press release, 26 November. Available at: **http://www.universityofcalifornia.edu/news/article/16864**.

Verspargen, B. (2006). University research, intellectual property rights and European innovation systems. *Journal of Economic Surveys*, **20**, 607–32.

WIPO (World Intellectual Property Organization) (2007). *What is Intellectual Property?* Available at: **http://www.wipo.int/about-ip/en/** (accessed 21 December 2007).

World Trade Organization (2007). *Intellectual Property: Protection and Enforcement*. Available at: **http://www.wto.org/english/thewto_e/whatis_e/tif_e/agrm7_e.htm**.

Yancey, A. and Stewart, C.N. (2007). Are university researchers at risk for patent infringement? *Nature Biotechnology*, **25**, 1225–8.

Ziman, J. (1996). Is science losing its objectivity? *Nature*, **382**, 751–4.

■ FURTHER READING

- Crespi, R.S. (2004). Patenting for the research scientist: an update. *Trends in Biotechnology*, **22**, 638–42. Aimed at research scientists, this article provides an introduction to the writing of a patent application. It explains the differences between scientific papers and patents.

- Seeber, F. (2007). Patent searches as a complement to literature searches in the life sciences – a 'how to' tutorial. *Nature Protocols*, **2**, 2418–28. This article provides easy to follow guidance on how to conduct web-based patent searches.

- Verspargen, B. (2006). University research, intellectual property rights and European innovation systems. *Journal of Economic Surveys*, **20**, 607–32. This article provides a comprehensive review of university patenting in the USA and Europe, drawing together a large selection of scholarly work on the topic from various disciplines.

- Loring, J. (2007). A patent challenge for human embryonic stem cells (**http://www.nature.com/stemcells/2007/0711/071108/full/stemcells.2007.113.html**). This commentary illustrates the fight, by an academic researcher, for the invalidation of a number of very broad patents that have the effect of stifling research and of preventing society from benefiting from the output of publicly funded research.

■ USEFUL WEB SITES

- The UK Intellectual Property Office: **http://www.ipo.gov.uk/**

- The European Patent Office: **http://www.epo.org/**

- The World Intellectual Property Organization: **http://www.wipo.int/**

- The United States Patent and Trademark Office: **http://www.uspto.gov/**

These web sites provide a wealth of information for those interested in intellectual property (introductory information, definitions, statistical data, current issues, legal texts, search facilities, etc.).

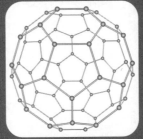

SECTION 2

Developing trends in scientists' communicating

. . . it is true that if we look at a glass of wine closely enough we see the entire universe. . . . If our small minds, for some convenience, divide this glass of wine, this universe, into parts—physics, biology, geology, astronomy, psychology, and so on—remember that nature does not know it! So let us put it back together, not forgetting ultimately what it is for. Let it give us one more final pleasure: drink it and forget it all!

Richard Feynman (1963) 'The relation of physics to other sciences'

2.1   **Science communication across disciplines,**
      by Joachim Schummer   53

2.2   **Communicating physics in the information age,**
      by Matthew Chalmers   67

2.1

Science communication across disciplines

Joachim Schummer

Introduction

Behind the general issues of communicating science to the public at large, there are the less obvious issues of communicating science to other scientists, although the challenges involved in each overlap. Indeed, because of the tremendous growth and increasing specialization of science and its fragmentation into numerous disciplines, subdisciplines and research fields, professional science communications are generally fully comprehensible only by experts in the same research field. The more distant the research fields are, the bigger the obstacles of understanding. Although all scientists might share some basic ideas about science, a scientist working in a far removed research field might meet obstacles of understanding about a research paper that are not totally different from the obstacles a well-educated non-scientist encounters.

However, the difficulties in cross-disciplinary communication result in many issues that are largely different from those associated with the detachment of non-scientists from science. Much more than non-scientists, scientists in their professional work depend on the best available knowledge in other fields. Moreover, if two or more scientists from different disciplines want to collaborate on a common project and do not understand each other, the project is unlikely to succeed. Effective cross-disciplinary communication is therefore a precondition for successful interdisciplinary work. Interdisciplinarity has been considered a source of innovation (see Weingart 2000), through cross-fertilization of disciplinary knowledge, which partly explains why cross-disciplinary communication issues have recently come to the fore. If each discipline or research field develops in isolation, science runs the risk of fragmentation—in stark contrast to the traditional notions of the unity of science. Because of the sheer scale of science and the difficulty of effective communication, it has long been impossible to achieve an 'overview' of all branches of science. Without such an overview, each field is more likely to go its own way, free of monitoring or control from outside. The risk then is that within each field experts take it upon themselves to decide on progress, in what are often publicly funded areas of research, because they are the only ones able to fully understand it.

This chapter provides an introduction to issues of cross-disciplinary communication and interdisciplinary research. The section that follows explains how and why these issues have historically emerged since the 19th century. Before addressing these key issues, we first need a better understanding of what disciplines are, how they differ from each other and how to describe their relationships. Because disciplines are both cognitive and social entities, there are cognitive as well as social strategies to improve cross-disciplinary communication, discussed in detail in later sections. Taking nanotechnology as the latest example to establish research across all major science and engineering disciplines, I then critically examine recent political efforts to support an interdisciplinary culture. Finally, I draw some conclusions about the extent to which cross-disciplinary communication can actively be controlled from the outside.

The growth and disciplinary fragmentation of science

As recently as the 18th century, different scientific disciplines in the proper meaning of the term hardly existed. Although different branches of knowledge of the natural world have been distinguished ever since antiquity, a scientist, then still called a natural philosopher, usually worked in many, if not all, areas, mostly as an amateur or member of a scientific society. Communication issues resulted from personal idiosyncrasies rather than from specialization. Scientific societies cultivated the exchange of ideas across the fields through convening regular meetings and publishing journals that combined all fields of scientific knowledge.

That situation changed drastically in 19th-century Europe, when newly structured universities became the institutions of formation of scientific disciplines. Formerly institutes of education for theology, medicine and law only, universities upgraded their faculties to offer not only undergraduate education but also higher degrees and doctorates in philosophy, which comprised everything we nowadays call natural sciences, social sciences and humanities. In addition, universities, which up until that time were merely educational institutes, established facilities in which professors and graduate students undertook research. The rapidly increasing societal demand for graduates enabled the philosophical faculties to flourish and grow, and to offer more specialized education and degrees. Professors, who formerly taught all kinds of courses, began to focus their teaching activities on fields related to their own research. As they found their own specialization, they wrote textbooks, started to edit specialized journals and trained graduate students to become professors of the next generation in their specific field. Increasing specialization thus led to the differentiation of disciplines, which defined and demarcated their own research fields, established communication structures and produced professionals and teaching staff. Simultaneously, the polytechnic schools typical of many European countries, founded in the late 18th and early 19th centuries to educate engineers, flourished and underwent a similar process of specialization and discipline formation. Here the original scope of mechanical, civil and military engineering was extended both by further specialization and in response to new industrial development and needs, to include new fields such as chemical and electrical engineering. The process of increasing specialization and fragmentation of science into disciplines, subdisciplines and research fields, which started in the 19th century, has continued throughout the 20th century up to the present day.

There are several reasons why science fragmented into increasing numbers of disciplines, subdisciplines and research fields. Most importantly, science has grown exponentially over the past two centuries, according to any quantitative measure, such as the number of publications, journals, scientists, research institutions and so on (de Solla Price 1961, 1963). All such numbers have roughly doubled about every 15 years during this period, which corresponds to an annual growth rate of some 5 per cent. It follows that, in order to keep up with the latest research, researchers need to focus on a comparatively small research field. When the field itself grows beyond the reading capacity of a researcher, the split into subfields seems unavoidable. Furthermore, science as a social activity depends on personal contacts with one's peers, in order to share implicit knowledge that cannot be communicated in written form and to distribute resources and career opportunities. As with any other social group, there is a limit to one's capacity to cultivate personal contacts, which determines the upper size of the group and thus fosters the splitting of the group once the limit is reached. And since the establishment of a new research field has become an important factor for the reputation of a scientist, this has provided further incentives for diversification and fragmentation.

Scientific findings are communicated primarily to peers or colleagues, which serves two main functions essential to modern science. First, results are shared among peers so that they can benefit from the results in their own research and, in return, give public credit to the authors. Second, peers might criticize a result as unsound or correct it in a subsequent publication, which keeps the methodological standard of the profession high and (at least in theory) prevents errors and inconsistencies. Both functions require that the communication is as precise as possible and fully comprehensible to peers. These requirements thus force scientists to formulate their communications according to the standards of their specific research field in a precise technical language, which fundamentally differs from ordinary language and the languages of other fields and disciplines. In other words, modern science requires that professional science communications are less comprehensible by non-scientists and, as science grows overall, by a growing fraction of scientists.

In addition to the long-term trend of the incremental fragmentation of science into specialities, there is a more recent trend over several decades towards increased interdisciplinary research (Gibbons *et al.* 1994). These two trends need not oppose each other, because interdisciplinary research frequently ends up in a new speciality or even in a new discipline, distinct from the disciplines it came from. The discipline of materials science, which emerged since the 1970s largely from physics and chemistry, is a recent prominent example. However, the trend towards interdisciplinary research is also driven by political and economic factors. Since public funding of research has grown much faster than the gross domestic product in most countries, societies generally require in return that research outputs have greater utility. In this regard, the two trends oppose each other. A new discipline or subdiscipline usually establishes itself by defining its specific research problems and foundations within the academic landscape, so that it is clearly distinguished from the other disciplines. In contrast, usefulness by general societal standards requires that research problems are coordinated across disciplines and geared towards societal needs. Thus, the politically desirable shift towards interdisciplinarity for the sake of improved usefulness, of which nanotechnology is the latest example, clearly acts against the long-term trend of the disciplinary fragmentation of science. Because fragmentation is a necessary outcome of the growth of science, tensions and issues of interdisciplinary communication are unavoidable, as we will see in the discussion of nanotechnology later in this chapter.

Disciplines and their relationships

In order to understand cross-disciplinary communication issues and the strategies to address them, we first need to have a better understanding of what disciplines are and how to describe their relations.[1]

The English term 'discipline' (from the Latin, *disciplina*) has a complex meaning, as the following sentence illustrates: Students (disciples) learn a certain doctrine (a discipline) by obeying strict (disciplinary) rules of a school (a discipline) and by practising self-control (discipline). A discipline is not simply an abstract set of information, but both a body of knowledge that is taught at a school and the social context of the school. Disciplinary knowledge requires a social context of transmission and education and a social body that reproduces itself by educating students to become future teachers. A scientific discipline thus comprises both cognitive and social aspects.

The *cognitive aspects* of a discipline refer to a body of knowledge of three kinds: concepts and beliefs, including facts, classifications, models and theories (knowledge of the world); methods for increasing and validating the knowledge of the world and for problem-solving (knowledge of methods); and values for judging the relevance and importance of the knowledge (knowledge of values). Hence, two disciplines differ not only in the specific set of information and concepts about the world, but also in what they consider important research questions, how to approach the problems and how to assess solutions. Cross-disciplinary communication issues thus arise not only because of different terminologies and information about the world, but also because of a different understanding of values and methods.

The *social aspects* of a discipline refer to a social body or a community of scientists who largely share the three kinds of knowledge and who feel committed to the community. The commitment includes active engagement in increasing and improving the disciplinary body of knowledge through research, in communicating it through publications and in teaching it to students. Like other social groups, a disciplinary community has rules for becoming a junior member (by graduation), for gathering (in society meetings), for

---

1. For more details, see the Further reading section at the end of the chapter, as well as Schummer (2004a) on which this and the following two sections draw.

distributing honour (through awards and society positions), for reproduction (through teaching appointments), for community-like behaviour (through codes of conduct) and for representing itself to publics. Being a member of a disciplinary community does not *per se* pose specific cross-disciplinary communication issues. However, the commitment to the community reinforces the cognitive issues and, because groups tend to stick together, it reduces the experience of cross-disciplinary communication (Box 1).

BOX 1 RELATIONSHIPS BETWEEN DISCIPLINES

The term *multidisciplinary* describes a loose or additive relation between the disciplines involved. For instance, a journal that compiles papers from many disciplines, like *Science* or *Nature*, is multidisciplinary as long as each paper is written by authors of the same discipline. In contrast, *interdisciplinarity* requires stronger ties, overlap or interaction between the disciplines involved. For instance, a paper or a research project is interdisciplinary if researchers from different disciplines successfully interact with their different disciplinary knowledge. Sometimes a paper is considered interdisciplinary if it does not exactly fit within a single disciplinary category. However, a discipline is not a static entity but develops flexibly over time—the category might therefore simply be outdated.

Two other terms are sometimes used in the sense of interdisciplinary, but, strictly speaking, they do not describe the relationship between disciplines. The term *cross-disciplinary* describes a move across the boundaries of disciplines, like that of information or communication. *Transdisciplinary*, although still a matter of intense debate, describes a form or state of science in which disciplinary structures, boundaries and commitments no longer exist. The term is frequently used to express political ideas of how science should or will be in the future, although project-based industrial research, which is usually far removed from academic disciplines and teaching, might already come close to that concept at times.

There are various models to describe the dynamics of disciplines and their interaction, but we are far from a full understanding. For instance, multidisciplinarity can be a preliminary step towards interdisciplinarity, such as when loosely aggregated disciplines begin to interact, but in most cases nothing follows. Similarly, many believe that interdisciplinarity is a preliminary step towards transdisciplinarity, such that all disciplinary boundaries vanish through intense interactions between the disciplines. However, interdisciplinarity can also inspire the mother disciplines to form new subdisciplines that each try to claim the new domain, like physical chemistry and chemical physics, biochemistry and molecular biology, and so on. Or, interdisciplinarity can result in a new discipline that grows independent from the mother disciplines, like materials science that emerged from the interaction between physics and chemistry.

Cognitive strategies for improving cross-disciplinary communication

Following this two-level definition of a discipline, we can distinguish between cognitive and social strategies for improving interdisciplinary communication, although these strategies can of course be interactive.

Cognitive strategies seek to level out the differences in disciplinary knowledge, or at least to enable successful communication despite the differences. Ideally, cross-disciplinary communication requires that scientists from different disciplines share the same knowledge basis—knowledge of the world, knowledge of methods and knowledge of values. However, as long as there are different disciplines in the proper sense, any common basis of overlap is modest in scale, because disciplines greatly differ in how they describe the world, in their methods for validating knowledge and solving problems and in assessing the quality and importance of pieces of knowledge. There are four cognitive strategies to smooth cross-disciplinary communication, but none of them provides easy solutions.

The most ambitious one is the philosophical idea of the unification of science through *reductionism*. In this approach, all the disciplines need to restructure their disciplinary knowledge such that it is translatable or reducible to some fundamental knowledge— the disciplinary knowledge of physics has generally been considered the ideal candidate. This requires at first that all descriptive knowledge of a discipline, i.e. specific concepts, models and theories, should be translatable into the language and theories of physics. More ambitiously, the methods and values of physics should also become the model for all the other disciplines. Popular as the idea was among 20th-century philosophers of physics, it turned out to be naïve because it disregarded the diversity of the cognitive structure of disciplines. However, the approach can smooth cross-disciplinary communication in very specific cases, if one finds a common knowledge basis into which fragments of knowledge from each discipline can be translated.

The second strategy, *simplification*, seeks a common basis in everyday knowledge. Because we share to some extent a common experience through language, a rich source of metaphors and images and a common-sense understanding of what matters and what is sound, this is a useful point to start with. However, there is a clear risk that over-simplification can create misunderstandings or the false impression of understanding where there is actually none. Over-simplification can also create artificial problems and solutions, suggested for instance by everyday metaphors, that can mislead rather than promote interdisciplinary research. Thus simplification can only be a preliminary step towards developing a more sophisticated form of communication. A second step, which Galison (1997) has found in his analysis of large-scale interdisciplinary research in particle physics, is the development of specialist jargons or creoles. These are inter-languages that are tailored to the needs of the specific project by combining necessary concepts from different languages and excluding unnecessary concepts that are mutually incompatible or incomprehensible.

A third strategy to smooth cross-disciplinary communication consists of reducing communication to restricted interfaces. *Modularization* divides up an interdisciplinary project into mono-disciplinary modules, i.e. subprojects, with well-defined types of information input and output from and to the other subprojects. However, modularization works only if the project architecture is simple, such that no disciplinary particularities affect the exchange of information, and if no unforeseen problems arise. Once such problems emerge, the interdisciplinary team may encounter particularly grave communication issues because of their inexperience in dealing with each other beyond the interfaces. Nonetheless, a well-designed organization of an interdisciplinary project in addition to regular exchanges and flexible arrangements can be a useful measure to avoid unnecessary communication problems.

The fourth strategy, *translation* or *mediation*, requires a translator who should ideally be educated in all the disciplines involved. Moreover, because there is no simple translation between the knowledge types of all disciplines, the translator needs to mediate not only between different types of description but also between different opinions on what is sound and important and on how to tackle a problem to best effect. Of course this requires sophisticated social and communicative skills, and it allows the mediator to control the project to a considerable degree. Mediation and translation would certainly be the best cognitive solution to cross-disciplinary communication issues, in particular because mediators can additionally educate scientists from different disciplines to understand each other better. However, mediators are rarely available because there is neither a profession nor a specific education for cross-disciplinary mediation, which leads us to social strategies for improving cross-disciplinary communication.

Social strategies for improving cross-disciplinary communication

Social strategies alone cannot directly enable cross-disciplinary communication because they cannot overcome the cognitive gaps between disciplines. However, they can establish social conditions under which mutual learning and understanding are improved and they can weaken the social commitments of scientists to their specific discipline to increase mutual openness.

The most ambitious approach is, of course, the establishment of broad multidisciplinary teaching rather than the monodisciplinary education that is characteristic of modern higher education. This would produce future researchers who have no disciplinary focus and commitment with the potential to freely communicate and collaborate with each other on any project. Yet, since the breadth of education within a limited period is achieved at the expense of depth, it is questionable if such education can qualify for cutting-edge research. We have seen that the fragmentation of knowledge into ever more specialized sub-areas has been driven by the growth of science; it cannot simply be reversed. However, a broad multidisciplinary teaching program could be useful to educate mediators who specialize in smoothing cross-disciplinary communication. If offered to all science students in their first year before disciplinary specialization, such a programme could also provide some basic understanding of other disciplines to improve future cross-disciplinary communication. Furthermore, specific teaching programmes that combine two or three disciplines can train students to work in corresponding interdisciplinary projects and settings.

Because education does not end with graduation, there are many opportunities to provide further education in disciplines other than one's own during a scientific career. Apart from formal further education programmes, there is a range of multidisciplinary science journals, popular magazines and books. Multidisciplinary journals, like *Science* and *Nature*, highlight important research in all disciplines, and thus provide ideal opportunities to keep up with what is going on beyond one's own discipline. Similarly, review articles summarize important developments and address a general scientific readership. There are also many multidisciplinary journals in specialized fields—sometimes wrongly called interdisciplinary—but which aim to prepare professional readers for interdisciplinary research and communication. Although popular science magazines and books are written primarily for non-scientists, they also attract a wide readership among scientists who seek an easy introduction to fields other than their own.

Apart from education, the social conditions for research can be considerably improved to foster the cross-disciplinary exchange of ideas. For instance, one can weaken the bureaucratic boundaries established between disciplinary departments, each with their own decision-making processes, budgets for research equipment, personnel, library, colloquia and so on. More ambitiously, a suitable architecture can make researchers from different disciplines close neighbours with formal and informal exchange on a regular basis rather than inhabitants of separate department buildings. From informal networks and consortia to interdepartmental research groups and centres with budgets and administration, research institutions can not only aim at interdisciplinary research results, but can also provide the social setting to facilitate and fashion cross-disciplinary communication. Finally, because money is always a powerful incentive, research grants that require interdisciplinary collaboration can help bring scientists from different disciplines together who would otherwise lack the opportunity to do so. On the bureaucratic side, this in turn requires that research funding agencies are not divided according to disciplinary lines.

All social strategies for improving cross-disciplinary communication and interdisciplinary research to some degree weaken the disciplinary identity and commitments of researchers. From a sociological point of view, it is questionable whether these group identities can be totally dissolved without substitution. On the other hand, establishing a new group identity can be a powerful means to abandon former allegiances. In order to replace a disciplinary group identity, one needs more than just a strong commitment to, say, a local research centre, because the new group identity needs to be established to be at least equivalent with the previous entity. Thus, the more powerful the establishment of a new group identity is, the more it employs the usual elements of discipline formation, which includes the establishment of all the cognitive and social components of a discipline already discussed. Furthermore, to make a new interdisciplinary field attractive, it needs to be popularized as particularly important, worthwhile enough to invest money in and capable of launching research careers. Typically propagators of a new field write histories which define the founders in order to shape its identity, and these generally refer to early and widely accepted authorities in order to add seriousness and attractiveness to the field.[2] Such histories provide a dynamic view of the field by placing current activities into the overall historical development and by providing extrapolations from the past to the future in the form of future visions. Expressed in simple terms with reference to general human needs, such visions provide quick answers to 'why questions', which researchers in highly specialized fields often have difficulty answering. By sharing the same visions and history, researchers originating from different fields can readily find a new common group identity.

2. Examples include Joseph Priestley's history of electricity from 1767, Wilhelm Ostwald's history of electrochemistry from 1896 and the historiographical and autobiographical efforts by James Watson in the 1960s to shape molecular biology (Abir-Am 1999).

The more powerful social strategies for improving cross-disciplinary communication are, the more they employ the classical elements of discipline formation. Of course, that runs the risk of resulting in a new discipline, which later inherits the same problems of cross-disciplinary communication. On the other hand, the weaker the strategies, the more likely their effects are to be no more than superficial and temporary.

The example of nanotechnology

Nanotechnology is the latest political effort to establish interdisciplinary research on a large scale, spanning all the established major science and engineering disciplines and acting in opposition to the long-term trend of disciplinary fragmentation. The potential practical applications of nanotechnology are considered enormous, as is its commercial significance. From about 2000, in all industrialized and many developing countries, huge national research programmes have been launched, making it an ideal field to study issues of interdisciplinary research and cross-disciplinary communication.

Nanotechnology is not simply a new specialized research field as the term might suggest. In fact, all definitions of nanotechnology are very vague. For instance, a widely used definition defines nanotechnology as the study of material structures in the scale of 1–100 nanometres (a nanometre being a billionth of a metre), in order to discover and exploit new properties of materials and devices that depend on nanoscale structures for useful applications. Indeed, almost all materials are structured in the nanoscale in such a way that the structure determines their properties. Chemistry has, at least since the mid-19th century, always complied with that definition, as for many decades have molecular biology, pharmacology, solid state physics, materials science and engineering, as well as larger branches of electrical, chemical and mechanical engineering and so on. While such vague definitions might be unsatisfactory from an academic point of view, in practice they allow the integration of many disciplinary research activities under the new umbrella of nanotechnology.

While the individual research activities that are nowadays called nanotechnology originate from many previous mono- and interdisciplinary research traditions, the umbrella concept of nanotechnology is a political idea. More specifically this idea was developed in the USA in the late 1990s—in part as a response to the 'Atom Technology Project' that the Japanese government had operated at its research institutes since 1993—and resulted in the launch of the US National Nanotechnology Initiative (NNI) in 2000. Because the NNI has been widely copied by other countries since then and is well documented, the following analysis focuses on the NNI and its impacts and discusses other countries only insofar as they substantially differ.

Science policy-makers have limited capacities to directly improve interdisciplinary research and cross-disciplinary communication, because their efforts are largely confined to social strategies, and in this example especially to funding. Lacking any direct impact on the cognitive structure of science, they can use only the power of words to convince scientists of the attractiveness, usefulness or necessity of cognitive integration. In this regard, the NNI started with the powerful idea of a revolution in science according to which the long-term fragmentation of disciplines would suddenly reverse towards a new convergence at the nanoscale. As the architect of the NNI, Mihail Roco, put it:

A revolution is occurring in science and technology. . . . At the nanoscale, physics, chemistry, biology, materials science, and engineering converge toward the same principles and tools. As a result, progress in nanoscience will have very far-reaching impact.

Roco and Bainbridge (2001, p. 1)

One might easily dismiss such a statement as naïve confusion of facts and wishes (Schummer 2008). However, it expresses the somewhat helpless vision of vanishing cognitive barriers between the disciplines (here, on the level of principles and methods), put forward in the hope that scientists might feel inspired to actually realize that vision. By combining the cognitive strategies of reductionism and simplification, the NNI and its forerunner organization promoted the idea that the world consists of simple Lego-like building blocks that can easily be imaged and rearranged (NSTC 1999). Once all disciplines agree on these building blocks, they would collaborate on rebuilding the world according to societal needs.

Strong efforts have been made to develop visions of a promising nanotechnology future to convince scientists of a common value basis that should both direct their collaborative research on the cognitive level and form a new group identity on the social level. In particular, the NNI, at least at the beginning, employed parts of Drexler's (1987) futuristic vision of 'molecular nanotechnology'; to bring unprecedented wealth, health and security through the development of robots on the molecular scale. They soon extended the scope of disciplines to also include computer, cognitive and some social sciences, all converging with nanotechnology in what has been called 'converging technologies', for the ultimate goal of enhancing the physical, mental and social capacities of humans (Roco and Bainbridge 2002). In addition to visions, the popularization of nanotechnology also comes with a standard history in which the US Nobel Prize winner Richard Feynman is posthumously made the founder of nanotechnology and an illustrious list of other Nobel laureates are described as early nanotechnologists.

Apart from the power of words, the NNI has distributed billions of dollars to support interdisciplinary research. The primary means has been the funding of interdisciplinary research projects and, particularly, centres at universities for a limited period, 'to provide strong support for the development of an interdisciplinary culture', as the reviewers of the NNI required (NRC 2002, p. 3). In addition, as an interagency institution, the NNI undermines the bureaucratic division of research funding agencies along outdated disciplinary lines, such as that between the National Science Foundation (NSF), which is responsible for physical sciences and engineering plus some chemistry, and the National Institutes of Health (NIH), which supports biomedical sciences and engineering plus some chemistry.

Compared to these strong efforts at developing an interdisciplinary research culture, the NNI and its sister initiatives in other countries have largely neglected education, the most promising social strategy for improving cross-disciplinary communication. For instance, in the USA the institutional support for developing interdisciplinary education only started in late 2004 with the establishment of a National Center for Learning and

Teaching in Nanoscale Science and Engineering (NCLT) that is modestly funded with less than 0.3 per cent of the NNI budget.

Fostered by nanotechnology research programmes world-wide, the global institutionalization of nanotechnology has gained an unprecedented momentum with annual growth rates of more than 50 per cent, such that after only a few years most major universities have at least one interdisciplinary nanotechnology centre or group (Schummer 2007). Since about 2004, a rapidly increasing number of universities have also offered undergraduate or graduate programmes in nanotechnology—particular in Europe, as a side-effect of the ongoing university reform that aims to make higher education compatible among European Union member states ('Bologna Process'), and in fast-developing countries such as China and South Korea, because new educational institutes and programmes are established according to current needs rather than to past models. In addition learned societies, and more recently commercial publishers, had launched more than two dozen nanotechnology journals by 2006, which are all intended to be interdisciplinary.

Whether the political impetus to create an interdisciplinary culture that smoothes cross-disciplinary communication is actually successful or not remains to be seen. In many regards, nanotechnology is still a loose multidisciplinary aggregation rather than interdisciplinary. Many of the supposedly interdisciplinary journals are still predominantly monodisciplinary, reflecting their allegiance to their publishers, like *Nano Letters* (published by the American Chemical Society), *Nanotechnology* (the Institute of Physics) and *Transactions on Nanotechnology* (the Institute of Electrical and Electronics Engineers). Beneath the surface, each classical discipline cultivates its own brand of nanotechnology, hence claiming a share in the huge budgets—and often demonstrating an affiliation to nanotechnology by adding the nano-label to ongoing research. Even if the journals are multidisciplinary, their individual papers are mostly written by authors from the same discipline (Schummer 2004b). A critical survey of 35 educational programmes in nanotechnology in North America, Europe and Australia has found that the vast majority of the programmes are monodisciplinary rather than interdisciplinary and that the nano-label is largely chosen to attract students to what are in effect traditional programmes (Brune *et al.* 2006).

Moreover, the fast institutionalization of nanotechnology research at universities does not necessarily result in a lasting interdisciplinary research culture. While the forms of institutionalization greatly differ from country to country, there seem to be two prevailing models, the 'temporary centre model' and the 'disciplinary model' (Schummer 2007).[3] Interdisciplinary centres, networks and consortia are temporary associations between researchers from different disciplines based on common interests. Because such associations are frequently decentralized institutions established for the acquisition of funding, it is uncertain whether they actually foster interdisciplinary research and cross-disciplinary communication and whether they continue to exist once the funding ends. In contrast, an interdisciplinary group consists of researchers from one or more

3. Among all countries Japan stands out because it has institutionalized nanotechnology largely at governmental research institutes through 5- and 10-year plans rather than fostering the academic institutionalization at universities. As a result the relative global institutionalization strength of Japan dropped from about 37 per cent in 1997 to some 7 per cent in 2006 (Schummer 2007).

disciplines who work on an interdisciplinary research project. If the group grows, it may upgrade to a department or school, which is more permanently integrated into the disciplinary structure of the university, and thus becomes the kernel of the formation of a new discipline. Through the funding policy of the NNI, the USA has a clear focus on the temporary centre model, whereas many European and Asian universities have already established nanotechnology at the department level and thus follow the disciplinary model. In both cases, however, interdisciplinarity may well prove little more than temporary. The political efforts made to cultivate interdisciplinarity thus navigate between the Scylla of a loose temporary aggregation and the Charybdis of a new discipline.

Given the complexity of the issue and the variety of cognitive and social strategies, the political efforts thus far appear to be very limited, short-sighted and sometimes naïve. The focus on funding is crippled by the short-term commitment of governments, such that nanotechnology might drop from the science policy agenda as suddenly as it appeared, whilst the most promising, albeit long-term, strategy of multidisciplinary education has hardly been tried; the same holds for cross-disciplinary mediation (for an exception, see Gorman *et al.* 2004). The substantial cognitive barriers between disciplines have been totally ignored or downplayed by propagating a simplistic Lego-like worldview that is hardly likely to convince established researchers (Schummer 2004a). And while the exaggerated visions may have generated some public interest, they have generally caused scepticism and divisions among some scientists. On the whole this suggests that nanotechnology might turn out to be a missed opportunity for improving cross-disciplinary communication.

Conclusion

The case of nanotechnology is instructive because it illustrates the complexity of interdisciplinarity and the problems of political control. Disciplines are both cognitive and social entities with their own dynamics of growth, change, fragmentation and mutual exchange. Substantial changes are measured in decades rather than in years and need to be supported by large sections of the scientific community rather than just imposed from the outside—otherwise the system reacts with pseudo-changes such as relabelling research to please science policy-makers. Because education is the very core of disciplines, any effective measure to control disciplinary dynamics and improve cross-disciplinary communication needs to start with education, which requires both willing scientists as teachers and patience to wait for the next generation. Although cross-disciplinary communication is a much desired goal, the sheer size and continuous growth of science requires that this can only be achieved either on a very general level or, if more detailed, for very specific interdisciplinary fields. However, if they find appropriate roles in research institutions, both generalists and interdisciplinary specialists can ease many important cross-disciplinary communication issues through mediation, organization and participation in research.

It would be naïve to consider disciplines themselves simply as an obstacle to cross-disciplinary communication, such that if they disappeared, communication would flow without boundaries. Disciplines are essential for education, for structuring knowledge and for controlling the internal quality of science; and they form the social dimensions of science that bear many of the characteristics of ordinary social life. As long as we have no fully functional substitutes for that, abolishing disciplines for the sake of cross-disciplinary communication would be throwing out the baby with the bathwater.

■ REFERENCES

Abir-Am, P.G. (1999). The first American and French commemorations in molecular biology. *Osiris*, **14**, 324–70.

Brune, H., Ernst, H., Grunwald, A., Grünwald, W., Hofmann, H., Krug, H., Janich, P., Mayor, M., Rathgeber, W., Schmid, G., Simon, U., Vogel, V. and Wyrwa, D. (2006). *Nanotechnology: Assessment and Perspectives*. Springer, Berlin.

Drexler, K.E. (1987). *Engines of Creation: the Coming Era of Nanotechnology*. Anchor Books, New York.

Galison, P. (1997). *Image and Logic: a Material Culture of Microphysics*. University of Chicago Press, Chicago.

Gibbons, M., Limoges, C., Nowotny, H., Schwartzman, S., Scott, P. and Trow, M. (1994). *The New Production of Knowledge: the Dynamics of Science and Research in Contemporary Societies*. Sage, London.

Gorman, M., Groves, J.F. and Shrager, J. (2004). Societal dimensions of nanotechnology as a trading zone. In: *Discovering the Nanoscale* (ed. D. Baird, A. Nordmann and J. Schummer), pp. 63–73. IOS Press, Amsterdam.

NRC (National Research Council) (2002). *Small Wonders, Endless Frontiers: a Review of the National Nanotechnology Initiative*. National Academy Press, Washington, DC.

NSTC (National Science and Technology Council) (1999). *Nanotechnology: Shaping the World Atom by Atom*. NSTC, Washington, DC.

Roco, M.C. and Bainbridge, W.S. (eds) (2001). *Societal Implications of Nanoscience and Nanotechnology*. Kluwer, Dordrecht.

Roco, M.C. and Bainbridge, W.S. (eds) (2002). *Converging Technologies for Improving Human Performance: Nanotechnology, Biotechnology, Information Technology and the Cognitive Science*. National Science Foundation, Arlington, VA.

Schummer, J. (2004a). Interdisciplinary issues of nanoscale research. In: *Discovering the Nanoscale* (ed. D. Baird, A. Nordmann and J. Schummer), pp. 9–20. IOS Press, Amsterdam.

Schummer, J. (2004b). Multidisciplinarity, interdisciplinarity, and patterns of research collaboration in nanoscience and nanotechnology. *Scientometrics*, **59**(3), 425–65.

Schummer, J. (2007). The global institutionalization of nanotechnology research. *Scientometrics*, **70**(3), 669–92.

Schummer, J. (2008, in press). From nano-convergence to NBIC-convergence. In: *Deliberating Future Technologies: Identity, Ethics, and Governance of Nanotechnology* (ed. S. Maasen, M. Kaiser, M. Kurath and C. Rehmann-Sutter). Springer, Berlin.

de Solla Price, D.J. (1961). *Science Since Babylon*. Yale University Press, New Haven, CT.

de Solla Price, D.J. (1963). *Little Science, Big Science*. Columbia University Press, New York.

Weingart, P. (2000). Interdisciplinarity: the paradoxical discourse. In: *Practising Interdisciplinarity* (ed. P. Weingart and N. Stehr), pp. 25–42. University of Toronto Press, Toronto.

■ **FURTHER READING**

- Klein, J.T. (1990). *Interdisciplinarity: History, Theory, and Practice*. Wayne State University Press, Detroit. This is a classic, and still the best, introduction to the concept of interdisciplinarity with a huge bibliography. More detailed and more recent books on interdisciplinarity by the same author include *Crossing Boundaries* (University Press of Virginia, 1996) and *Humanities, Culture and Interdisciplinarity* (State University of New York Press, 2005).

- Kline, St. J. (1995). *Conceptual Foundations for Multidisciplinary Thinking*. Stanford University Press, Stanford, CA. This book analyses methodological and conceptual issues of interdisciplinarity from a systems theory approach. The fourth part of the book, where the author analyses various fallacies in creating interdisciplinarity, is particularly useful.

- Weingart, P. and Stehr, N. (eds) (2000). *Practising Interdisciplinarity*. University of Toronto Press, Toronto. This anthology investigates the recent shift towards interdisciplinarity from a science studies perspective. The theoretical analyses are complemented by many useful case studies on the actual practice of interdisciplinary research centres and funding institutions worldwide.

- Schummer, J. and Baird, D. (eds) (2006). *Nanotechnology Challenges: Implications for Philosophy, Ethics and Society*. World Scientific Publishing, Singapore. This volume—written by philosophers, historians and sociologists of science—provides an introduction to the recent emergence of nanotechnology across the disciplines and its various philosophical, ethical and societal issues.

■ **USEFUL WEB SITES**

- **US National Nanotechnology Initiative** (NNI): **http://www.nano.gov/**. Under 'Resources/ Publications' this web site provides useful and detailed documentation concerning the US Nanotechnology Initiative since 1999.

- **Nanoforum.org**: **http://www.nanoforum.org/**. This UK-based web site provides the most comprehensive information on EU activities in nanotechnology. For full access to all sources, free registration is required.

- **Nanotechnology Now**: **http://www.nanotech-now.com/**. Among the numerous nanotechnology news portals this is probably the most comprehensive one, with the typical mixture of research, business and visionary enthusiasm.

- **Rethinking Interdisciplinarity**: **http://www.interdisciplines.org/interdisciplinarity**. The French project 'Interdisciplines' has organized a number of international online conferences and seminars, including one on Rethinking Interdisciplinarity in 2003/2004 with full papers available.

- **Bibliography on Interdisciplinarity**: **http://www.grad.washington.edu/Acad/interdisc_network/ ID_Docs/bibliography_Interdisc.pdf**. Although a bit outdated (2003), this is currently (2008) the best available online bibliography on interdisciplinarity compiled by Gail Lee Dubrow from the University of Washington.

2.2

Communicating physics in the information age

Matthew Chalmers

It is intriguing to note that modern science emerged shortly after Gutenberg's invention of the printing press some 550 years ago. Some historians take the view that when Nicolas Copernicus presented his heliocentric and heretical view of the universe in 1543, he was able to do so only because he had access to a range of printed sources that enabled him to compare and contrast older ideas (Eisenstein 1983). Such materials, the story goes, allowed Tycho Brahe and others to pick up where Copernicus left off, free of prejudice and armed with precise mathematical tables that would prove vital in driving the scientific revolution forward.

Whether or not this version of history is correct, few would deny that the transition from manuscripts to print had a profound effect on all forms of scholarly commun-ication. Printing allowed new knowledge to be disseminated more widely, eventually generating a need among a growing number of proto-scientists (then termed natural philosophers) to stake claim to what we now call 'intellectual property' (see Schulze, Chapter 1.3 this volume for a discussion). By the end of the 17th century, a handful of academic journals—including the still-running *Philosophical Transactions of the Royal Society*—were in place.

Today, the act of producing persuasive claims to new scientific knowledge, evaluating them and making them available to others remains at the core of physics and all other sciences. Moreover, it has enabled a publishing industry worth several billion pounds per year to spring up around it. So far, popular media—in particular radio and television—do not seem to have had much impact on the professional practice of science, although they have undoubtedly broadened science's public audience. But the same cannot be said of the next sea-change in the way humans communicate: the internet.

In just a few decades, the internet has made it commonplace for scientists to publish and download information 'at the click of a mouse' (see Gartner, Chapter 3.2 this volume), not to mention making e-mail by far the preferred medium for academic correspondence. Physicists like to consider themselves ahead of the other sciences in this regard, for good reason. The familiar face of the internet—the World Wide Web—was invented by UK physicist Tim Berners-Lee in 1990 while working at the European Laboratory for Particle Physics, CERN (Organisation Européenne pour la Recherche Nucléaire), near Geneva, from where the idea spread rapidly (Berners-Lee 1999). In the following year, this allowed US particle physicist Paul Ginsparg to establish what future historians may come to view as the first nail in the coffin of the traditional publishing model: the arXiv preprint server, a freely accessible data base that at the time of writing (2008) contains almost half a million papers from physics and closely related disciplines (Box 1).

BOX 1 ARXIV AND PREPRINTS

arXiv.org (http://arxiv.org/) is an online repository for scientific papers (i.e. preprints) that have not been published in a commercial or other scientific journal. Preprints speed up the dissemination of knowledge and particle physicists in particular have a strong tradition of circulating them that dates back to the mid-1950s, when such papers were exclusively in print form. The particle physics laboratory CERN, and later SLAC (Stanford Linear Accelerator Center) in the USA, later formalized this process and established a preprint catalogue. With the arrival of the World Wide Web in 1990, it was natural to put this catalogue online, leading to the term 'e-prints'. This feat was achieved in 1991 by theoretical physicist Paul Ginsparg then at Los Alamos National Laboratory, allowing a researcher to simply upload a paper.

Since 2001, arXiv—pronounced 'archive' since the 'X' represents the Greek letter 'chi'—has been hosted by Cornell University. It now contains preprints not just in theoretical particle physics, as it did originally, but in mathematics, computer science, quantitative biology and statistics, and has become phenomenally popular. The numbers of papers submitted to the site has increased steadily since it began, currently (2008) being added to at a rate of roughly 5000 per month. The site offers free access to a significant fraction of all the latest results in physics (CERN researchers, for instance, enjoy access to some 90% of the literature in their field), and gave birth to the open-access publishing movement. Although an advisory board ensures basic standards, the preprints are not peer-reviewed so users have to exercise their own judgement when assessing the quality of a paper. Since 2004, however, those who do not already have a preprint on arXiv.org require an endorsement from somebody who does. The purpose of this was to 'ensure content that is relevant and of interest to current research'.

While doubtless causing headaches for commercial journal publishers, the idea of using a web site as a convenient virtual library from which to access information is only the beginning of what the web has to offer. Two particular applications are highlighted in this chapter. Online diaries in the form of blogs (short for web-logs) provide a simple, asynchronous and interactive communication channel that simply did not exist a decade ago. Sites such as Wikipedia, which allow users to continually re-edit and re-post content—in effect a form of collaborative authoring—are beginning to challenge traditional views of established knowledge and how it is communicated. This chapter will argue that, so far at least, physicists do not appear to be embracing the new Web 2.0 internet era as quickly as they did the original 'click and download' web as one of their preferred modes of communication.

Challenging the traditional model

Editors of major traditional print journals such as *Nature* and *Science* started to embrace the potential of the web in the mid-1990s, for example by digitizing production processes and offering quicker publication times, space for supplementary material and extensive search facilities. This has certainly improved access to the scientific literature for those who already subscribe to the journals. But as Gartner's chapter in this volume (Chapter 3.2) testifies, the underlying pay-for-access model of traditional journals looks increasingly outmoded in the internet age—which is hardly surprising, given that it grew up in an era when the most efficient way to disseminate information was to print and distribute a hardcopy journal.

The potential of the web to make scientific research, much of which is publicly funded, available to anyone who wants to access it is called open access publishing (see Gartner, Chapter 3.2 this volume for discussion). As is widely recognized, this is now causing the biggest source of upheaval in scientific publishing, provoking often heated debate between proponents (usually scientists) and commercial publishers (see, e.g., Voss and Enderby 2007). Typically, open access journals transfer the costs of peer review and other quality-control and production measures from the reader to the author.

Titles of this type have taken off in the biological and medical sciences, with sites like BioMed Central and the Public Library of Science hosting a large number of open-access journals. Recently, BioMed Central launched a physics arm called PhysMath Central, and in total there are currently (2008) about 35 open-access journals in physics—including the popular *Optics Express* and *New Journal of Physics* (*NJP*), which are now well established. As with traditional physics publishers such as Institute of Physics Publishing (which publishes *NJP*), PhysMath Central recognizes the importance of arXiv in the physics community by enabling researchers to submit papers to it directly from arXiv, or to submit papers to both sites simultaneously.

The take-up of open access journals by physicists has been slower than in the biological and medical sciences, where presumably the literature has a much broader appeal; this may be in part due to the popularity of arXiv. However, CERN is spearheading a project called 'sponsoring consortium for open access publishing in particle physics' (SCOAP) that would see all results in particle physics published in open access journals.[1]

Shaping knowledge online

The rise of online versions of traditional journals, preprint servers and open access titles illustrates how the web is extending access to scientific knowledge to those who can log on to the internet. Indeed, as more electronic papers appear, the web is offering new approaches to quantifying the impact of that knowledge. Over 100 electronic paper data

---

1. http://www.scoap3.org/.

bases such as Google Scholar and Scopus are now well established, and physicists have been among the first to devise improved productivity measures based on citation analysis (Box 2).

BOX 2 PRODUCTIVITY MEASURES

Online data bases of scientific papers offer powerful new ways to quantify the merits of scientific research or even scientists themselves. One such measure developed in 2005 is the h-index, named after its inventor physicist Jorge Hirsch, which has become a popular way to rank individual scientists (Hirsch 2005). A researcher has an 'h-index' of 10, say, if he or she has published 10 papers that have attracted 10 citations. According to Hirsch, after 20 years a 'successful scientist' might have an h-index of 20, an 'outstanding scientist' an h-index of 40, and a 'truly unique individual an h-index of 60.

In 2006, physicist Michael Banks modified the h-index to assess individual subject areas.[2] A topic with an 'h-b index' of 10 means that there are at least 10 papers on that topic, each of which has been cited at least 10 times: since some topics have been around longer than others, this number is then divided by the number of years that papers on that topic have been published in order to assess how important a particular topic is currently.

Yet another measure, developed the same year by physicist José Solerby, ranks a researcher's 'creativity'.[3] A paper that has lots of references but only a few citations has a low creativity index, C_a, while a paper with just a few references and lots of citations will have a high creativity. The creativity of a particular scientist can then be calculated by summing the total creativity for every paper that author has written. None of these measures would have been practical before the advent of searchable, electronic data bases of scientific papers.

Perhaps a more profound issue is the extent to which the web might shape scientific knowledge itself. In principle, the scientific paper could change radically when set free from its print format. For example, by relaxing the constraints of space that print journals have to adopt for practical reasons, authors could use less jargon and fewer acronyms and therefore increase the potential audience of the paper—and perhaps in doing so stimulate interdisciplinary research (see Schummer, Chapter 2.1 this volume). Perhaps such moves would also generate greater opportunities to describe blind alleys and flawed trials, thereby preventing other researchers from unproductive use of their time; the inexorable increase in computer memory allows whole data sets to be included alongside a paper, making results easier for others to verify, refute or develop further.

So far, this has not taken off in physics. While some papers do contain links to additional images and simulations, as well as links to supplementary graphs and methods, a typical *Physical Review Letters* paper looks much the same today on-screen as it did on paper 50 years ago. Similarly, the advantages of including raw data sets and similar material do not seem to be attractive to researchers as yet. The web, it seems, has the potential to shape the knowledge that physics generates, but in practice that potential awaits exploitation.

2. arXiv:physics/0604216v2; this form of citation is routinely adopted for arXiv—see
   http://arxiv.org/help/arxiv_identifier.
3. arXiv:physics/0608006v1.

This same might be said of 'open peer review', a system that promises to make the world of peer review more transparent. Open peer review (see Wager, Chapter 4.1 this volume for discussion) strives towards this goal by ensuring that a paper (once it has undergone some basic level of scrutiny) spends a certain amount of time online for anybody to comment on before being peer reviewed in the conventional fashion by appointed referees. This approach would be impracticable without the web, although so far it does not seem to have captured the imagination of the physics community. To date only one journal in the physical sciences—*Atmospheric Chemistry and Physics*—has employed the open peer review model. When it comes to tampering with the self-regulating institution of science, it seems that the adoption of new technology is inevitably approached with some caution.

The interactive web

The World Wide Web is quickly becoming a much more social medium than the one-way 'click and download' model that ensures that researchers can readily access scientific information online. The bundle of applications referred to as web 2.0 encourages people not just to use the web as a reference source but also to interact with it by contributing content, and through automated systems that make intelligent connections between related online material. The eventual aim, which Berners-Lee had in mind from its beginnings (Berners-Lee 1999, pp.191–215), is to make it easier for people to create and share content, ranging from digital photos of their cats to entries about the paradoxes of quantum mechanics in user-edited encyclopaedias.

Chief among these emerging forms are blogs, which are so far the most widespread manifestation of user-generated content on the web. A blog is essentially an online diary that features regular entries or 'posts' by one person or a small group, typically listed in reverse chronological order, and archived. Postings on a blog can include text (including documents), images, audio, web video and links to other web sites, including other blogs. If the blogger chooses to allow it, anyone reading the blog can add their own comments.

Since the term was first coined in 1997, blogs have become an internet phenomenon. At present, 2008, over 110 million are listed in the blog directory *Technorati*[4] and this number increases at the almost unbelievable rate of 175,000 per day. Of course, the number of blogs created does not necessarily relate to the number that are maintained by regular posting; many blogs are simply abandoned. With so many people voicing their opinions, the content often has limited appeal. But blogs have also been lauded as a new form of 'citizen journalism', for example providing us with eyewitness accounts of dramatic events such as the terrorist attacks of 11 September 2001 and mobilizing a network of support following the Indian Ocean tsunami on Boxing Day 2004 (Allan 2006). Blogs have also sparked a phenomenon called 'blooks'—books that have been inspired by blogs—although this is yet to take off in the cyberworld of physics.

4. http://technorati.com/.

Several science magazines and academic journals have set up blogs, featuring rolling reports from conferences or updates on the latest science news. However, these do not generate anything like the level of interaction, i.e. the number of comments, enjoyed by the blogs of many individual scientists. The most obvious reason is that blogs are seen by their users—known collectively as the blogosphere—as the antithesis of traditional media, freed from the established constraints and conventions of established publishing and emerging from a counterculture that runs deep among web enthusiasts. Indeed, an increasing number of physicist bloggers are firing shots across the bow of science journalism, using their blogs as a forum to debunk what they perceive as inaccurate or misleading science reporting in a range of news media.

Cosmic Variance[5] is currently one of the most popular physics blogs, with some 3000 readers per day. The blog hosts discussion 'threads' ranging from the latest preprints in theoretical particle physics to the marriage vows of one of its authors. This romantic thread illustrates the versatility of blogs: 'science' blogs do not have to be about just science and 'citizen' blogs can include science. *Cosmic Variance* is written by seven physicists and cosmologists including Sean Carroll, with the occasional post from a 'guest blogger', and recently started selling its own branded clothing. But most physics blogs—there are about 100 listed in the 'aggregator' site *Mixed States*[6]—are maintained by individual researchers and updated three or four times per week.

Blogging for physics

While some see them as vanity projects, physics blogs are starting to have an impact on the way researchers communicate—both with each other and with the outside world— at least among a subset of physicists. For instance, papers published on arXiv are now able to cite blog entries via 'trackbacks', which are a crucial feature of web 2.0. While the original hyperlinks' of the web are one-way, simply pointing from one site to another, trackbacks are a means of notifying a web page that another web page has added a link to it. In the case of arXiv, if a physicist writes about a particular paper in their blog and sends a trackback request, that blog entry is then automatically included on the paper's page.

Such a combination of arXiv preprints and blog comments has already proved fruitful. In 2006 physicists Robert Alicki, Daniel Lidar and Paolo Zanardi produced a revised version of their paper on quantum error correction (arXiv:quant-ph/0506201v2) in the light of discussions a few months earlier on Dave Bacon's blog *The Quantum Pontiff*.[7] And in late 2007, US cosmologist Lawrence Krauss was forced to alter the conclusion of a preprint after coverage in *New Scientist* and the *Telegraph* newspaper—which was duly dissected in the blogosphere—exposed a misleading claim in his paper about the fate of the universe (arXiv:0711.1821v2). Once he had altered his conclusion, Krauss—who is one of

5. http://cosmicvariance.com/.

6. http://mixedstates.somethingsimilar.com/.

7. http://scienceblogs.com/pontiff/.

the most 'media savvy' physicists around—popped up on the relevant blogs to apologize for making misleading claims that were then reported in news media, and apparently also managed to correct his quotes in the online version of the *Telegraph* article.

What is clear is that physics blogs have huge potential as forums for genuine scientific debate—another recent example has been the many assessments of 'surfer dude' physicist Garrett Lisi's arXiv preprint (arXiv:0711.0770v1) presenting 'an exceptionally simple theory of everything'—a cyberspace discussion picked up by the mainstream news media. However, the role of blogs in shaping science is not always as clear-cut as these few examples suggest; discussion threads can easily be derailed by people promoting their own 'off-the-wall' theories, political opinions or even personal grudges. Of course, such behaviours are as old as science itself, but the blogosphere makes it much easier for them to be disseminated and debated, at least on the web. In an attempt to keep such distracting elements out of arXiv's system, all trackbacks (which are generated for only a small minority of preprints) are moderated by its eight-member advisory board.

With some physicist's blogs offering 'plain English' explanations of the latest physics stories, blogs are proving themselves to be useful for communicating with the wider world. Indeed, *Cosmic Variance* has propelled Carroll to someone who the mainstream news media now regularly call on for accessible explanations of difficult science. More importantly perhaps, the informal diary style of blogs can be an effective way to give non-scientist readers an idea of what it's like to actually do science. In the *Quantum Diaries* project, for instance, 33 physicists blogged about their life and work to celebrate the World Year of Physics in 2005; a similar but smaller project has just been set up by US particle physicists keen to promote the USA's role in CERN's new particle accelerator, the Large Hadron Collider (LHC).[8]

Such authentic 'warts and all' accounts of science are not easy to convey within the constraints of mainstream academic journals, where convention usually demands the use of rhetorical devices—such as writing in the passive voice and removing any evidence of the author's personal involvement—in order to connote a sense of disinterestedness and objectivity. The hard news values and space restrictions of newspapers or even popular science magazines also leave little room for conveying the underlying motivation for or challenges involved in a piece of research, with the exception of the occasional longer feature articles.

For those who work in mainstream news media, especially on popular science magazines, blogs have also become a potential source of news (see Allan 2009 for discussion). Bloggers are often aware of new research results—possibly because they are their own—long before any journalist or university press officer is able to go public, and some take the time to post footage of talks from conferences that a journalist (or indeed other researchers) might be unable to attend.

Blogs can also be highly effective rumour mills; presumably scientists are much more comfortable posting a (possibly anonymous) comment on a blog than they are calling up an editor with a tip-off. A recent case in point was speculation on particle-theorist Peter Woit's popular anti-string theory blog *Not Even Wrong* about possible delays in the hotly anticipated start-up of CERN's LHC, in the form of a series of 'updates' such as:

8. http://www.uslhc.us/.

From the comments here and e-mail I'm getting, it appears that others are hearing these same rumors: the first physics runs are likely to be in 2009, not 2008, due to problems that have shown up as they have started cooling down some sectors of the machine.[9]

In part by way of response to this and other (mostly inaccurate) rumours, the subsequent issue of CERN's internal newsletter carried an article by CERN Director general Robert Aymar that began:

In an age of blogs there are seemingly no secrets, so by the time Lyn Evans [LHC project leader] gave his talk on the status of LHC commissioning on 13 September, everyone seemed to know about plug-in modules, beam position monitors and transmitters embedded in ping-pong balls.[10]

Eventually, this led to an article in *Physics World*, which started:

Robert Aymar, the director general of CERN, has dispelled rumours that a series of buckled electrical connectors at the Large Hadron Collider will delay the accelerator's official start-up date of May 2008.

Cartwright (2007)

As this modest episode shows, neither traditional media nor those at the top levels of scientific management any longer have an exclusive position in setting the science news agenda.

Physics bloggers and journalists head to head

The extent to which journalists should report blog postings as news is a moot point. The fact is they can, and the greater the potential for blogs to provide hooks into a good story, the more journalists will monitor them. Staying with the theme of particle physics (one of the most popular topics in the physics blogosphere), possible sightings of the Higgs boson is a topic that has recently received widespread coverage in the press. Finding this fundamental particle is one of the main goals of the LHC. But journalists are after their own prize: the Higgs exclusive, which would likely turn out to be the physics scoop of the decade. When these two media forms—blogs and traditional news media—overlap, as will happen increasingly often once the LHC starts taking data in the next few years, sparks can fly, and that can make a good story in itself.

In January 2007, particle physicist Tomasso Dorigo mentioned the Higgs particle on his blog *A Quantum Diaries Survivor*[11] in connection with a small 'bump' in data taken at the US particle physics laboratory Fermilab. This led to a two-page news story in *New Scientist* immodestly entitled *Glimpses of the God particle* (Ananthaswamy 2007), which

---

9. http://www.math.columbia.edu/~woit/wordpress/?p=600.

10. http://bulletin.cern.ch/eng/articles.php?bullno=41/2007&base=art#Article2.

11. http://dorigo.wordpress.com/.

was soon picked up by *The Economist* (Anon 2007) and *Wired* (Borland 2007) all of which helped engender even greater publicity in the blogosphere.

This work had at that stage yet to be initially reviewed by the 500-strong experimental collaboration of which Dorigo was part, which left him rather exposed. He then tried to distance himself from the story by claiming on his blog that the lack of statistical significance of his bump—which he said he had taken care to present faithfully in his original posting—was not properly conveyed by the *New Scientist* journalist. He even cut and pasted the e-mail exchange he'd had with the journalist to support his case.[12]

In the end, Dorigo came under heavy fire from his collaborators,[13] who have a strict system in place for releasing results to the wider world, and the episode resulted in the collaboration issuing a formal list of guidelines for its blogger members, extracts from which state:

Before Blogs, CDF [the experimental collaboration of which Dorigo was a member] was able to choose the forum where this information would be presented in public first. Bloggers [are now able to] to shortcut this norm by posting the result immediately . . . We propose that Bloggers kindly refrain from posting discussions on [such results] until the collaboration has officially presented them first. We consider our meetings and 'business' discussions to be private.

The Higgs episode demonstrates how blogs could become channels for establishing priorities—the lifeblood of science—without having to wait for those discoveries to be published or verified by others. Of course, such claims would then be verified by peers for the priority to mean anything. Some physicists already seem to use arXiv for this purpose by posting papers that have been submitted to peer-reviewed journals—then telling journalists that they are not allowed to cover the paper until it comes out in the refereed journal. From the point of view of media law, however, anything that is placed online is considered an act of publishing.

Traditional media outlets do, however, need to be wary of announcements made in blogs, particularly if they are working with a single source. But such caution is already well established as good professional practice, whatever the medium. As the 'cold fusion' episode illustrated (see Lewenstein 1995 for discussion), working outside the conventions of scientific verification can be a dangerous and ultimately self-defeating course. Because of its newsworthiness, *New Scientist* was justified in covering the Higgs story as picked up from Dorigo's blog and from another *bona fide* particle physicist John Conway's postings at *Cosmic Variance*.[14] But wherever they appear, too many stories of this nature—which are going to become increasingly difficult for journalists to resist as rivalry between the main LHC experiments, as well as with Fermilab, hots up in the next few years—could severely dilute the impact of an actual Higgs discovery and risk a situation where 'media fatigue' sets in.

12. http://dorigo.wordpress.com/2007/03/01/the-mssm-higgs-signal-buried-in-my-plot/.

13. http://dorigo.wordpress.com/2007/03/09/the-trouble-with-talking-about-physics/.

14. http://cosmicvariance.com/2007/01/26/bump-hunting-part-1/.

Blogging as reporting

The differences between blogging and journalism reveal some of the downsides of blogs as reliable online information sources. For example, there is no onus on bloggers to be balanced in what they say or to seek opinion from both sides of an argument. And while bloggers quote heavily from mainstream media stories, press releases and other blog postings, they rarely generate new and potentially valuable insight to a story by quoting well-regarded academics who are not members of the blogosphere. Of course, in many cases, the blogger is an expert in that subject, but this shortcoming of blogs as news providers is neatly illustrated by my own actions on picking up the rumours about a delay in the LHC schedule on Woit's blog in late 2007: as a member of the editorial staff on *Physics World* magazine, I was obliged to follow a more traditional communication practice—to pick up the phone and call my key contact at CERN to find out what was going on.

Another key difference between bloggers and science journalists is the issue of agendas. A reader of a reputable magazine or newspaper can be fairly confident that the journalist's agenda is primarily to further his or her career by writing accurate, balanced and informed stories, with more opaque 'opinion' traditionally relegated to the comment pages in an attempt to lend objectivity to the rest of the content. With blogs, there is no easy way of knowing whether the author has an axe to grind.

Most physics blogs also have very narrow audiences of people, often with similar interests and levels of knowledge, which can make the technical level of posts patchy and often inappropriate for the majority of outsiders. Indeed, the audience of many blogs is intentionally small. But a more worrying aspect of the blogosphere as far as providing reliable information goes is that the vast majority of people who read and post comments on blogs also have blogs of their own—about 80% according to a small poll carried out by *Physics World* in 2006 (Griffiths 2007). This can create an artificial sense of importance and controversy around a subject, with contributors writing self-servingly of things seen on other blogs—it's worth noting that maintaining a high-quality blog can take up a serious fraction of a researcher's day job.

Wikifying the web

A few years ago the idea that one of the most-consulted sources of information in the world would be an online encyclopaedia that can be modified by anyone who uses it would have seemed ridiculous. But that is exactly what has happened with *Wikipedia*, probably the most well known of the web 2.0 sites. As it has turned out, the self-correction that is built in to the system has worked pretty well—barring occasional controversies, erroneous modifications are usually quickly corrected by another user. Indeed, a study carried out by *Nature* in 2005 suggested that, for science, *Wikipedia* is almost as accurate as the *Encyclopaedia Britannica*, with the average *Wikipedia* entry containing around four inaccuracies to *Britannica's* three (Giles 2005).

From the Big Bang to quantum computing, there is a wealth of physics information on *Wikipedia*, though detail is sparse on more obscure topics. Most physicists appear happy to use *Wikipedia* as a quick reference—as an initial source to be checked, generally as a prelude to more detailed and systematic research. The level of trust physicists have in the encyclopaedia varies markedly, however. Some physicists who responded to a *Physics World* survey, such as Robert Helling of Jacobs University, Bremen think that the quality of articles in *Wikipedia* is high 'especially in sufficiently general topics that have been edited, verified and refined by a large number of editors'. Other respondents remain unconvinced; 'I wouldn't dream of reading *Wikipedia* for physics' says Nobel laureate Philip Anderson, and 'nor would I trust it if I did' (quoted in Griffiths 2007).

In fact, *Wikipedia* is the most successful example of the general concept of a user-editable web page, or 'wiki', which is being put to use by physicists for other purposes. In a large collaboration, a wiki can be a great way to get round the problem of disseminating tacit knowledge and 'tricks of the trade', especially when collaborators are spread over the world and cannot take part in coffee-room discussions—important if under-researched aspects of 'informal' science communication. Such wikis already exist in particle physics, where they are gradually evolving into comprehensive repositories of information for large experimental collaborations such as ATLAS at CERN.[15] But beyond such specialisms, use of wikis is as yet limited; according to the *Physics World* survey only 14% of those surveyed had contributed to a work-related wiki (Griffiths 2007).

Socializing online

As suggested earlier, despite being at the forefront of the development of the web in the 1990s, physicists have been slow to embrace some of the innovations offered by web 2.0, committed enthusiasts apart. The same could be said of 'social tagging', a form of classification by web 2.0 users widely used in sites such as the photo-sharing forum *flickr*. Users choose 'tags' to describe their photos—'me', 'London', 'red' and so on—and can then search on not only their own tags, but also those of millions of other users.

This bottom-up form of classification can also be used for scientific information. Indeed, the sites *Connotea* and *CiteULike* were set up specifically to apply social tagging to scientific papers. The idea is that when you find a useful paper online, you save it in your account and add tags describing the content of the paper. If this was just a way for individuals to manage their references, it would not be terribly exciting. The innovative aspect of sites such as *Connotea* is the social dimension; you can see what other users are saving and search their tags to look for new papers on a topic of your choice. But in practice, biologists seem to have adopted the concept more eagerly than physicists—with 'genetics' and 'metabolism' being among the most popular tags on *Connotea*; however, only one respondent to the *Physics World* survey said they had used such sites (Griffiths 2007).

---

15. https://twiki.cern.ch/twiki/bin/view/Atlas/WebHome.

In the wider world, there are a number of immensely popular web 2.0 destinations. *MySpace* is currently one of the most prominent—a 'social networking' site that allows each of its users—a staggering 200 million by early 2008, mostly teenagers—to produce a personal homepage with photos and details of their likes and dislikes. The page also prominently displays how many 'friends' the user has on the site, and friends can add comments to each other's pages, creating something of an online popularity contest. Although many physicists have signed up to *MySpace*—for example, string theorist and popular-science author Michio Kaku is currently (2008) listed with an impressive 11,931 friends—rival site *Facebook* (which has accumulated in excess of 55 million users since its creation by a 20-year-old Harvard student in 2004), is the more likely target for academics, given its origins. (Initially membership was limited to those at Harvard, then gradually expanding to include anyone with an e-mail address ending in .edu or .ac.uk. Currently (2008) anyone with a valid e-mail address can join.)

Facebook is certainly popular among physicists at CERN, but only in terms of socializing —not as a professional communication tool. There could, however, be a more serious use of social networks for physicists. Jennifer Golbeck, a computer scientist at the University of Maryland, has found that a social network like *MySpace* contains useful information about who knows whom and how much people trust each of their contacts. Golbeck is working on algorithms to use this information to tell you how much you should trust someone you do not know, based on their position in your social network; this trust measure could then be used by scientists and their collaborators. For example, researchers might allow access to an early version of a paper only to people whose trust rating is above a certain threshold.[16] More pragmatically, there is evidence that the world of science is beginning to appreciate the political potential of social networking sites; in the funding crisis of early 2008 relating to UK astronomy and particle physics, several new *Facebook* groups emerged. Though initially with rather few registered names, the clear intent was to galvanize the physics community into more effective and coordinated political lobbying to reverse threatened shortfalls in funding.[17]

It is clear that a brave new world of blogs and trackbacks, wiki peer review, social tagging and trust networks awaits physicists who dare to venture into web 2.0. But this inter-active usage of the web may have to wait until the *MySpace* generation hits the labs before it fully make its mark on physics. There are some signs that user-generated video could be one of the first to do so. For example, a lecture about atoms and heat that was posted on the hugely successful video-sharing site *YouTube*[18] in September 2007 was viewed some 90,000 times in less than 3 months, and a search for 'physics' among the site's enormous data base returns nearly 40,000 results. Indeed, the University of California at Berkeley has started to post whole lecture courses on *YouTube*, while *SciVee*[19]—which has been dubbed a 'YouTube for science'—allows researchers to upload videos of themselves discussing their papers.

---

16. See Golbeck's research page at: http://www.cs.umd.edu/~golbeck/publications.shtml. The paper in question is at: http://www.cs.umd.edu/~golbeck/downloads/networkStudy.pdf.

17. For example, http://www.facebook.com/group.php?gid=6772194284.

18. http://www.youtube.com/.

19. http://www.scivee.tv/.

Surviving the information age

It is easy to forget that we are still witnessing the early days of the web as a medium for science communication, and every business that has paper and printing at its roots is grappling with how best to harness its potential. There is good reason to be cautious. For example, the increasing ease of access to scientific knowledge could lead to spurious contributions that actually amplify the need for traditional quality control measures to be maintained, or new ones introduced. While some physics blogs may rise above the noise to provide a useful communication channel between researchers and publics, and occasionally as forums for genuine scientific debate in ways that would otherwise be very difficult, they are unlikely to ever replace magazines such as *New Scientist* or *Physics World* which, by operating according to longstanding professional norms, provide consistently high-quality (though non-peer-reviewed) content for broad audiences. As such, they occupy the important space between peer-reviewed academic journals and news media outlets. Furthermore, the impact of the internet is likely to vary markedly between different subfields of physics. Particle physicists, for example, have been circulating pre-prints for 50 years, and despite continuing to publish in peer-reviewed journals, arXiv is now home to the vast majority of papers in the field.

As this chapter has testified, it would be short-sighted for any physicist to underestimate the importance of the transition from print to online media. The rapid development of the web is often hailed as an information revolution. But it is better viewed as the latest technological leap in an evolution that began with the development of writing some 5000 years ago and against more recent innovations such as the Gutenberg press. Science as we know it is a relatively recent chapter in this story of human creativity, and it has certainly flourished in, and to a large extent become defined by, the medium of print. How much the web will change science from now on is a question only answerable with hindsight. That is, of course, if future historians are able to read the myriad of data formats in which the information age is currently being written.

Acknowledgement

Substantial parts of this chapter originally appeared in *Physics World* in January 2007: 'A revolution in bits' (written by the author) and 'Talking physics in the social web' by Martin Griffiths, now a parliamentary advisor for physical sciences, IT and communication.

■ **REFERENCES**

Allan, S. (2006). *Online News—Journalism and the Internet*. Open University Press, Maidenhead.

Allan, S. (2009). Making science newsworthy; exploring the conventions of science journalism. In: *Investigating Science Communication in the Information Age: Implications for Public Engagement and Popular Media* (ed. R. Holliman, E. Whitelegg, E. Scanlon, S. Smidt and J. Thomas). Oxford University Press, Oxford.

Ananthaswamy, A. (2007). Higgs boson; glimpses of the God particle. *New Scientist*, 2 March 2007, 8–11.

Anon (2007). Higgs may fly. *The Economist*, 8 March 2007.

Berners-Lee, T. (1999). *Weaving the Web: the Original Design and Ultimate Destiny of the World Wide Web by its Inventor*. HarperCollins, New York.

Borland, J. (18 June 2007). Rumors in physics blogosphere test faith in 'God particle'. *Wired*. Available at: **http://www.wired.com/science/discoveries/news/2007/06/higgsboson**.

Cartwright, J. (2007). CERN boss quashes delay rumours. *Physics World*, **20**(11), 8.

Eisenstein, E. (1983). *The Printing Revolution in Early Modern Europe*. Cambridge University Press, Cambridge.

Giles, J. (2005). Internet encyclopaedias go head to head. *Nature*, **438**, 900–1.

Griffiths, M. (2007). Talking physics in the social web. *Physics World*, **20**(1), 24–8.

Hirsch, J. (2005). An index to quantify an individual's scientific research output. *Proceedings of the National Academy of Sciences*, **102**(46), 16569–72.

Lewenstein, B.V. (1995). From fax to facts—communication in the cold-fusion saga. *Social Studies of Science*, **25**(3), 403–36.

Voss, R. and Enderby, J. (2007). The open-access debate. *Physics World*, **20**(1), 22–3.

■ FURTHER READING

- Berners-Lee, T. (1999). *Weaving the Web: the Original Design and Ultimate Destiny of the World Wide Web by its Inventor*. HarperCollins, New York. Written by the physicist behind its development—now director of the World Wide Web consortium—this is an account of how the web was invented, discussing the history and philosophy behind the internet.

■ USEFUL WEB SITES

- **Bloggers for peer reviewed research reporting: http://bpr3.org/.** This blog is dedicated to discussions about peer-reviewed scientific research.

- **North Carolina science blogging conference: http://scienceblogging.com/.** This site relates to the North Carolina science blogging conference, at the time of writing (2008), in its 2nd year. It has postings on many issues such as the role of blogs as science journalism outlets, outreach, etc., although overwhelmingly from the point of view of bloggers themselves.

- **Mixed States: http://mixedstates.somethingsimilar.com/.** This is the 'mixed states' blog which helpfully lists all the main physics blogs.

SECTION 3

Accessing contemporary science

For also knowledge itself is power.

Francis Bacon (1597) *Meditationes Sacrae*

3.1 **Science and the online world: realities and issues for discussion,**
by Scott L. Montgomery   83

3.2 **From print to online: developments in access to scientific
information,** *by Richard Gartner*   98

3.1

Science and the online world: realities and issues for discussion

Scott L. Montgomery

Introduction

As a body of knowledge, science exists because its practitioners record, preserve and exchange both words and images. Science is deeply a matter of communication. The forms that scientists have available to them for sharing their work inevitably have a profound importance and influence—what physically can and cannot be expressed, how it can be shaped, what counts as acceptable information and analysis and, not least, who has access to it. Knowledge, in short, is inseparable from its forms. This means that new forms of communication create new possibilities for scientific knowledge, not just as a corpus of data and interpretations but also as a means of doing work, establishing expert communities and interacting with 'outsider' publics. New modes of communication create new forms of science and ways of being a scientist.

The matter does not end here. Communication is involved in other aspects of science, its socio-political dimension, its economics and, more generally, its cultural presence. For much of the 19th and 20th centuries, professional scientific knowledge was restricted to a set of literary containers, mainly journals, books, reports and theses, to which non-experts had limited access. Such removal helped underwrite the elevated status of the scientist-expert. It also provided support for the grant-in-aid contract system of funding technical work, through the promise of a true knowledge-product (e.g. the article), as well as a system of measurement of significance via numbers of publications and citations. Peer review, the 'publish or perish' syndrome, the epistemological and economic power of the technical journal—all these have roots in modes of communication developed in the 17th and 18th centuries, refined and expanded in the 19th and 20th centuries, and now, in the early decades of the new millennium, undergoing stress and change in the online world.

Whether the internet is a truly revolutionary force in science cannot yet be said with any finality. We are, as it were, too immersed in everything. Nonetheless, it is already evident that the work and substance of science, as well as the politics of technical knowledge, are being altered, not least by the vastly expanded access to this knowledge. Culturally speaking, scientists have not lost their elite rank among intellectual producers —indeed, wider dissemination of technical information has helped reinforce this prestige in some ways (scientists are still treated as embodiments of their knowledge). But greater public access means more immediate response from 'outside,' thus more ambiguous boundaries between science and non-science, and, perhaps above all, greater account-ability for scientists.

Historical context

Shifts in human communication

Online science is part of a shift in human communication. What sort of shift is this, at a fundamental level? Valuable answers can be found by considering, even briefly, the historical evolution of how understanding of the natural world has been inscribed and transferred.

Physical recording of knowledge began with pictures. Cave paintings dated to between 30,000 and 40,000 years ago, which appear to correlate with long-term settlement in pro-tected valleys, reveal close observation of animal physiology and, to some degree, behaviour. These 'documents' were dependent upon technology—locating, preparing and applying certain metal oxides and charcoal, using human saliva or animal fat as applicants. They were also immobile and possibly secret, the realm of their actual exchange extremely limited. The next major step involved the evolution of writing, as marked by cuneiform ('wedges' pressed into clay tablets), sometime between 4000 and 3000 BC. With this new capability, coming from the first great city-states of Mesopotamia, recorded knowledge became portable and mobile. Cuneiform relied on technological innovation: tablets and styluses, methods of baking (in kilns). Though derived from pictures, it did not easily include them, being used for record keeping (payments, receipts, taxes). Yet this is significant: in its oldest manifestation, writing met a need to document concrete knowledge-events.

By late Roman times, communication had undergone two more transformations. The first came with the scroll during the rise of Nile Valley civilization in the third millennium BC; the second involving the shift from scroll to codex (Latin for 'block of wood'), the earliest form of the book, in the first few centuries AD, and related to the spread of Christianity. Each transformation relied upon, but also brought, a combination of social, technological and epistemological changes. Scroll culture, evolving with the commun-ication needs of Egyptian and Roman society, expanded written knowledge to include literary, religious, philosophical and scientific material, and came to erect an industry on a gigantic scale (Roberts and Skeat 1983). The codex, meanwhile—flat leaves, or 'folios', bound between covers to create a book—was adapted by the emerging Christian church possibly as a means to best preserve the letters of the Apostles (Pagels 1992). Manuscript culture, from the 5th century onward, made things more portable still, rendering the acts of reading and writing more simple, and over time, particularly from the 12th century onward (when Graeco-Arabic science entered Europe via translation), re-installing the power of images—the 'art' of medieval texts.

Such changes gained dynamism with the introduction of moveable type. The printing press as a vehicle of intellectual transformation in the West is by now a well-known concept (Eisenstein 1980; Johns 2000). In purely physical terms it is the least impressive of the innovations we have noted, yet the book finally made recorded knowledge, earlier restricted to a few intellectual communities, available on a vast new scale. Books multiplied knowledge enormously, urging the change from an oral to a written society, expanding the audiences for education and study, stimulating not only readership but authorship, inquiry and exchange, therefore laying the groundwork for modern science itself. By the Renaissance, images came to be as important as text in certain epochal works—Vesalius' *De fabrica humani corporis* (1543), for example, setting the standard for modern anatomy. Over time, the printed book gave rise to differently scaled versions of itself—the monograph, article, atlas and so on. A single basic form could be adapted to an ever-expanding array of intellectual needs.

Internet science: the overall context

How, then, does all this relate to internet science? We see in past transformations that each evolutionary step arose from social and technological circumstances and influenced these in turn. Each expanded the mobility of knowledge, thus its capacity to be shared. Each increased the span of knowledge-producers and consumers, and the communities of use. And each episode multiplied the contents of knowledge itself, as well the forms in which it occurred.

All of this appears to be happening today. The internet arose from post-World War II Cold War society, from Big Science and Technology. The origins of the internet were with civilian scientists working for the US Defense Advanced Research Projects Agency, or DARPA (founded in 1958, in direct response to the launch of the first satellite—Sputnik— by the Soviet Union in October 1957). But its true flowering during the 1990s coincided with other socio-cultural developments, notably the shift to a knowledge-based society in advanced nations, the spread of computers and also the phenomenon of globalization.

The internet marks a new stage in electronic communication, which began well over a century ago with the telegraph. Its advance, we should be clear—like all advances in technology—is not deterministic, compelling society to evolve in certain predetermined directions. Instead, as historians of technology would stress, the internet has been shaped by how people—individuals, institutions, companies, governments—have chosen to employ it. In David Nye's (2006, p. 61) terse phrasing, the internet, like all technology, 'is an extension of human lives'.

Electronic communication has a material basis. It cannot exist without computers, electricity and literacy. In 2007, about 5.7 billion people did not own a computer, 1.6 billion had no access to power and 774 million adults (two-thirds of them women) could not read or write. Even accounting for overlap, we can be assured that over 2 billion people remain shut off from the online community entirely. Thus, to speak of the internet as 'universal' in any sense is at best naïve. The 'digital divide' also means a global science and technology divide and everything this suggests.

What new capabilities has the online world brought to science, meanwhile? At first it joined the immediacy of the telegraph with the substance of the book, but then went on to include the entire panoply of digital information—video, photography, audio, animation and many types of specialist communication (satellite imaging, spectrography, tomography, etc.). Digital transfer expands the domain of electronic substance, from static text and images, to dynamic visuals, auditory elements and combinations of all these. Science relies profoundly on all of these areas, and it thus stands to benefit more from the online world than perhaps any other realm of knowledge-production.

There are complexities. The internet also brings a larger audience for science, changes in author–reader relations and new modes of interaction. Many online journals now provide chances for readers (normally peers, but not always) to comment directly on published material. There has been a decided trend to also open up online science to public comment—whether in the style of online forums, official blogs, feedback sections or some other means. Such commentary exposes technical work to direct scrutiny and thus involves a new level of accountability. Debates have raged about whether this is appropriate or advisable. Either way, it can be counted as a strategic move, clearly done with the hope of encouraging appreciation and endorsement more than outrage or dismissal. For science, the internet is a many-splendoured tool in the politics of knowledge.

Advantages and disadvantages for online science

Advantages

What specific benefits has the online world brought for science, in a proven sense? There are many, in fact, some yet to be fully understood. When the internet was in its infancy during the 1960s, its first connections were between universities, and the most popular application was e-mail. This should not be surprising, but is significant. E-science greatly expands the interaction among researchers, first and foremost. The immediacy of this contact and the ability to combine it with the sending of information—whether raw data, final text or graphics, supporting documents and so on—means that collaboration has many new possibilities. Researchers in different cities or nations can collaborate far more easily, without meeting face-to-face. New research communities can form rapidly around new findings or ideas, or gather around the fire of a political issue—government censorship, scientific fraud or peer review are recent examples. E-science creates new opportunities for coming together, communicating, thus also generating articles and other publications.

As I have already noted, the online world also provides new interaction between scientists and other audiences. This includes the public at large, certainly, but also more specialist publics. What is called 'outreach' has been enormously, almost (for some topics) overwhelmingly, expanded. Research centres, government agencies, large companies and even important journals now routinely have online 'news of the profession' sites, as well as educational resources and comment opportunities on technical subjects (*Nature's*

'Climate Feedback' site is a case in point). This is no less true for trade associations, like the World Nuclear Association, and non-governmental organizations (NGOs) such as Greenpeace. Individual scientists, too, through personal web sites, proffer their own work and opinions, and encourage commentary in return. The internet also opens up science to itself, providing new avenues for specialists to keep up with their own fields and to engage with, possibly even enter into, other disciplines of interest (for example, a field botanist drawn to climate change participates in a forum, makes contact with researchers in the field and collaborates on new studies). At the same time, online encyclopaedias like Wikipedia make a certain level of scientific research in every discipline integral to the larger realm of freely available knowledge. The internet thus has this dual potency of making science more public, more interactive with society, and making the public more of a 'consumer' of science.

Obviously, there are limits here. The degree of online interactivity between scientists and non-scientists remains restricted. Scientists still tend to see 'outreach' as a subset of public relations. The desire to recruit support is all the more urgent in some nations, such as the USA, where a resurgence of challenges to the value and neutrality of science have surfaced in the 2000s, via debates over evolution, climate change and stem-cell research in particular. To a significant degree, the internet has given scientists and scientific organizations a potent new tool in the unending struggle over 'hearts and minds'. Yet it is no less a weapon for their opponents.

Organizations such as the Royal Society and the American Association for the Advancement of Science (AAAS), as well as the national science/engineering academies, clearly recognize that the higher levels of exposure and accountability brought by the internet in democratic societies also impose increased responsibility. More than ever, it is viewed as part of the labour of science to represent itself accurately, honestly and continuously and to frame a voice for science on social issues. What about the contents of research itself, then? Here it makes sense to list some of the advantages that are now apparent:

• The internet allows data to be made available immediately and without interruption. The transfer of this information is more rapid than in the past, usually more efficient, cheap and thorough. The quantity, as well as the mobility, of technical information is greatly increased.

• Data can be shared simultaneously among any group of collaborators, making it possible for individuals at different institutions to work together on a project at any time. This includes the sharing of background literature as well as experimental results.

• The capabilities of computer graphics, coupled to the internet, expand the visual dimension to science in many ways, allowing for the use of colour, 3D and 4D modelling, animation and more. This has meant that new types of information, previously unavailable for publication, can now be included in articles, journals or reports.

• Electronic journals have many practical benefits: multi-reader access, reduced storage space, ease of searching/indexing and independence from library/institution hours, among others. They also provide entirely new capabilities of great use, such as links in a reference list (articles thus become 'local libraries' on a specific subject).

• Online articles and reports can be produced and published more quickly, with a shorter turn-around time, a major consideration in frontier areas. This includes more facile submission, peer review, revision and final publication. It also means a far wider potential distribution.

• Depending on the field, articles and other publications are often posted for comment to the larger research community (more peers). Online publications can then be revised, over time, or else left up with commentary (peer review 'exposed').

• Important talks, presentations and lectures are now often recorded, stored and made available. This includes material from conferences, research symposia, department seminars, invited lectures and also courses—all of which can aid scientists in their daily work, as researchers and teachers.

• Over time, certain mundane (but still core) aspects of science have been rendered more efficient by the internet: the search for funding sources; grant applications; procurement; storage and filing of documentation and other bureaucratic tasks.

Disadvantages and limitations

A number of limiting factors and possible drawbacks have also become apparent for online science. These are not trivial. They are fundamentally related to the vastly increased exposure of scientific information and accessibility to it. One of the primary concerns is with intellectual property rights, both in terms of copyright for published material and for ownership of proprietary data. Can researchers put articles they have authored on their own web sites? To what degree can graphics and other material from published research be copied and used in lectures? Who owns the data that appear in an online report: the provider or the publisher? These and other, related questions have yet to be fully worked out in many cases—there are no clear, generic standards or rules. Researchers in a number of disciplines—biotechnology, materials science, molecular genetics, high-energy physics—have found themselves at the heated intersection of competing claims over ownership, duplication rights and even access (see Schulze, Chapter 1.3 this volume). The ease with which material can be copied, modified and redistributed online introduces a new set of concerns. Journals have found themselves just as much in the business of finding ways to restrict access as to explore methods of expanding it.

Plagiarism and fraud are other enhanced risks. The digital world is fluid and dynamic. Such dynamism vastly overflows the ability to enforce the rules established to govern the culture of print. As every teacher knows, the internet provides enormous opportunity for copying and using the work of others. More serious is the possibility of altering original research results, faking data or deleting unwanted information that might cast doubt on an important finding. When coupled with the intense competition of contemporary science and the spectacular awards that come from breakthroughs, such realities may become seductive. Thus, there is an added burden with e-science, to find ways to track and discourage such abrogations of trust.

Online science holds out other limiting possibilities. Problems can arise if information or data are 'leaked' prior to actual publication—that is, shared with other researchers, who then (unknowingly or not) make it more widely available. The same problem of potential 'pirated versions' applies to proprietary data of any kind and to early drafts of articles on controversial topics (e.g. public health-related studies). It is nearly a cinematic cliché at this point that the portability of technical information on the internet renders it vulnerable in particular ways. It can be stolen, lost, corrupted (by viruses, hackers) or erased. It can also become the source for dangerous uses—related to biohazards or nuclear technology, for example. Indeed, this is true in general: the world-wide nature of the web makes any unclassified information, e.g. on how to build a simple atomic bomb or a polio virus, available to anyone with a computer and internet connection. Data, and other significant information (of many kinds), can vanish or be displaced when web sites are abandoned, whether for economic or bureaucratic reasons, an extremely common phenomenon.

There are social aspects to consider as well. Increased exposure to non-scientific scrutiny also brings the possibility of increased politicization. Not all scientists are comfortable with having to act as the public face of their disciplines. It remains difficult for many to realize that the internet is a gigantic, semi-organized, but always *public* realm, where almost anything that is done on it can be made available for millions to see. It can work as a rumour mill; thus reputations can be quickly and deeply damaged, or else raised by false information. In controversial areas, with political and ideological implications, it is possible for a relatively small number of individuals, by amplifying their opinions through various outlets (multiple web sites, blogs, chats, feedback) to cast doubt on the value of an entire discipline—exactly what occurred with regard to climate change science in the USA. The culture of modern research is one of proprietary claims, internal critique and interpretive community. The culture of the online world is one of instant exchange, externalization and open contact. It is not clear that these two domains are entirely in concert as yet.

One last possible drawback to e-science is the potential for elements of entertainment to be used to make science 'sexier' or 'more fun' in order to better compete for popular attention. The perils of popularization—selective pandering to public tastes and anxieties; favouritism granted to spectacle and 'wonders'; reliance on stereotypes; use of kitsch—have been often written about and delineated. It is not clear that the internet forces scientists in any way to adopt these techniques.

Balance sheet: is the internet really good for science?

Considering all factors, online science holds far more positive promise than disadvantage. Too many of the benefits discussed and listed represent real advances in the conduct of scientific work and its ability to be expanded in nutritive ways. Like the codex and the printed book, the internet multiplies communication itself—how it is done, whom it involves and what it contains.

Indeed, the internet is now indivisible from science and constitutes an essential element in its growth. This does not mean the internet should be considered a 'natural force' with the power to compel future directions in research or to make all knowledge universal and free. E-science defines a new realm that has carried forward intellectual and material traditions of the past, while modifying them and allowing for alternatives, all within the framework of conventional decision-making. It would be naïve to expect a new technology to enter into the mainstream of intellectual work and there wash away, as if by divine gesture, all existing social relations and dependencies, including commercial ones. The internet is decidedly a boon to science, yet its specific capabilities will take time to explore, using experiment, innovation and (*mirabile dictu*) accommodation.

Major forms of e-science

Perhaps the most alluring aspect to e-science is the multiplication of forms in which knowledge can occur. During the late 19th and 20th centuries, scientific communication grew from a small number of literary forms—mainly books, articles and letters in the early 1800s—to include theses, monographs, editorials, essays, book reviews, news reports, commentaries and more by the 1970s and 1980s. This overall expansion, limited by being bound to the printed word, took more than 150 years to achieve. In less than a tenth this time, a greater array of new forms has arisen for e-science. What are some of these forms? The following provides a basic list.

E-periodicals

This includes the online versions of major technical journals and their archive of back issues, as well as online-only journals, plus newsletters and other 'lighter' publications. The ease with which material can be submitted, edited and put on the web has helped encourage many journals to add new commentary sections ('public discussion') and to create online-only subsidiary periodicals, for example specializing in one area, thus increasing the range of publication and authorship. E-periodicals have continued to grow in number and kind.

Preprint and e-print archives

Preprint archives, where early versions of articles as well as other material (editorials, essays, etc.) are made freely available, exist for some fields, and there is much desire to extend them to others. The oldest, arXiv, covering physics, mathematics and computer science, dates from 1991 and is counted a great success, with well over 1 million entries in 2007. Open-access initiatives also exist for geology, biology and biomedicine, via the Copernicus Publications, BioMed Central and Public Library of Science sites. Open access in general has been a battleground between researchers and publishers, and is discussed further below. E-print archives, on the other hand, are online collections of articles, presentations, lectures and other material that an organization makes available, often for free and on a specific topic.

Professional web sites

A host of resources, including publication outlets, are made available by scientific organizations and groups. These include professional societies and associations, research institutes and also industry groups, NGOs, international agencies and think-tanks. Not all such organizations provide information for use by researchers, though many do, either through available data bases, journal links and searches or online reports and documents.

Government web sites

With the exception of defence-related and other national security data, a great deal of information is regularly, if somewhat variably, provided through the web sites of government entities. Such entities include ministries (health, space, agriculture, etc.) and major agencies, but also legislative committees (e.g. energy and natural resources), important bureaus and other official departments. Democratic governments, in particular, have now put many types of information in the public domain—periodicals, reports, data archives, data bases, all types of images, bibliographic resources and software too. There is also news about major research efforts and about budgets. Finally (but this is hardly a complete list) and critically, the research funding process, involving grant application and awards, is now conducted almost exclusively online.

International web sites

Included here are online resources provided by various global organizations, such as the European Union (EU), the World Bank, the World Health Organization (WHO) and programmes of the United Nations (UN). The UN, in particular, offers a great deal of open-access information about food and agriculture, water resources, energy, demographics, environmental issues and through the Intergovernmental Panel on Climate Change (IPCC).

Institutional web sites

In this category are to be counted the sites for libraries, museums, hospitals, universities and institutes or major charities (e.g. the Wellcome Trust, the Gates Foundation). This is a diverse group, and the resources they provide for scientists range from the trivial (brochures) to the essential. University libraries, in particular, act as guardians of access to online journals in every field, plus tools to help researchers in their work. Universities also maintain web sites for individual courses that can be helpful to scientists looking for guides in their own teaching.

Web logs (blogs)

Science blogs are now legion and offer a new kind of expressive outlet for scientists, a 'watering hole' for exchanging views. There are gateway sites for such blogs (e.g. http://scienceblogs.com/), which add an extra layer of commentary but serve a valuable function. Blogs in general are a significant addition to the literature of science. They provide 'soap boxes' for personal expression, forums for comment by scientists and non-scientists alike, and places where all aspects of science are open for discussion, from research problems to economics and politics to philosophy of science: in short, they are sites where science's larger place in society, as reality *and* as subject of inquiry, is given its due.

Personal web sites

Personal web pages also allow scientists great freedom of expression, technical and otherwise. Researchers use such sites in many ways, but above all to distribute their views and work. Published articles or early drafts are often put up. Other professional communications, like book chapters, speeches, expert testimony and slide presentations are also often given. Personal web sites can be 'look at me' catalogues, too, thus giving interesting revelations of the competitive side of technical work.

Audio-visual resources

Online science can also take advantage of the internet's ability to transmit audio and visual information. Podcasts are commonly used to imitate radio, but are capable of much more than this and seem likely to evolve further (see Redfern, Chapter 5.3 this volume). Often undervalued is the video universe (e.g. *YouTube*), which holds out enormous potential. Video recordings of conferences, lectures, debates, documentaries and more are already available. But other possibilities of science performance abound: demonstration (experiments), education (training, workshops), personal statement and popularization.

This brief list of resources and possibilities, incomplete as it is, should suffice to underline a crucial fact. E-science is not only more diversified and disseminated than p-science (print), but is also more ecumenical and more social. The online world not only opens science to wider participation and to new expressive forms, but by doing so it shows that the boundaries once assumed to divide scientific endeavour, *in its essence*, from other domains such as politics, economics and the wider contemporary culture, are largely mirages. The technical material of any discipline and exchanges among scientists over this material remain largely inaccessible to non-experts (including scientists in other fields), true enough. Yet through such activities as blogging, e-mailing, personal web pages, online feedback and commentary, the demand for open access journals and other such public expressions and exposures, researchers reveal themselves as intellectuals deeply embedded in the socio-cultural moment.

Major issues for contemporary science communication

All evidence points to the conclusion that the internet represents a new, more advanced stage in the long evolution of human communication and that this stage has enormous, if complex and not yet fully understood, benefits for science. The question therefore becomes: what sorts of issues and implications for science does this technology hold? An introductory discussion of this question would include the following.

Political/cultural dimensions

We have discussed how the online world 'opens up' science in certain ways. But it does the opposite too, in a wholly concrete manner. Indeed, the ultimate effects of e-science are showing themselves to be diverse and not always predictable. Two contrasting trends make this evident.

First, e-science deepens certain inequities in the access to knowledge. The digital divide, noted earlier, creates 'scientific wealth' and 'scientific poverty.' Technical know-how is obviously a powerful engine for development generally; thus nations, or portions of them, lacking access to such know-how (as a subset, perhaps, to a lack of educational opportunities as a whole) are effectively falling further behind each year from what they could have achieved. Policies by governments to prevent, retard or generally constrain such access thus reduce the real world possibilities for development. Regimes that seek firm control over online access, and thus e-science as well, show that they view knowledge and communication as primarily political capital. Cases in a number of countries (China, Burma, Iran) indicate that it remains wholly possible to limit such access, and to use the internet as an effective means to track dissidents, including scientists.

Where access is expanded, the internet can greatly stimulate the growth of scientific work (thus science and technology wealth) in developing states. E-science creates many possibilities for collaboration between scientists in the global 'North' and the 'South', at many levels (research, fieldwork, publication, consulting, teaching, conferences), with decided benefits for both sides. Online work of this type has proven to be of enormous benefit to the analysis and treatment of disease, for example, as well as food and water problems in Africa and South Asia. This increase in the international dimension to science helps lay the groundwork for technology transfer as a critical aspect to economic progress. The internet, in short, today acts to enhance the role of science as an ambassador not merely of goodwill but good works in the struggle against poverty and its effects.

Another related issue has been the spread of English as an international language. English, as is well known, serves as the *lingua franca* of science. Thus, the shift of science to the internet brings with it a burden for non-English parts of the world, a linguistic divide that may deepen the digital divide. Important questions abound: who controls the learning of English in poorer nations? Is it reserved for traditional elites, through the established education system, or made more broadly available? Does the advent of English create a pathway for Western views and lifestyles to displace local culture? Or does it rather provide a medium for individuals to open new doors of opportunity, for scientists especially to become more global citizens, gain access to much of the world's technical literature, participate in international conferences, publication and help raise the level of science within a home nation?

E-science, in short, brings new possibilities and challenges—not least to authoritarian governments who seek to manage communication. It pays to be blunt in this sphere: the online world isn't universal, free or removed from traditional means of control. Inasmuch as it expands human contact, it also expands the options for both expression and suppression. In much of the world, as yet, the internet has not rendered science more powerful, politically speaking, nor more democratic.

Epistemological issues

Within the advanced world, e-science, we have said, is larger and more diverse in its forms than p-science could ever hope to be—but what might this mean in terms of knowledge itself and its direction? This is a question that cannot be answered easily or fully at this stage. However, a few comments, worthy of discussion, may be made:

• The capabilities of the internet have greatly aided, even urged, interdisciplinary scientific work. This has helped new subdisciplines to flourish and find a firm place within the larger domain of science. New interpretive communities of this kind are legitimized more rapidly than in the past, by online journals and other intellectual infrastructure.

• The overall impact of these trends is to make science at once more specialized and more cross-fertilizing. During the past few decades, major fields—biology, geology, astronomy, etc.—have been redefined as overarching categories: life sciences; earth and space sciences; etc. The intermingling of subspecialities in these categories is itself a measure of progress.

• New kinds of multi-disciplinary work have also begun, involving domains previously considered external to science. Already expanding in some universities are the fields of 'sustainability studies' and 'climate studies', for example, which include not only 'hard' sciences *per se*, but also economics, sociology, research on the media and policy analysis.

• In the realm of expression, specifically, e-science is having two important effects. The first of these is to reinvigorate the place of the written word at the centre of knowledge (internet science multiplies writing). The second effect is to elevate the power of imagery. This involves new forms of evidence, data and analysis, all based on greatly expanded capabilities for display, and that greatly heighten the aesthetic appeal of technical work, possibly with some confusions too.

• An unresolved question is whether e-science can improve the degree of 'scientific literacy' among publics, including children. If so, what levels are acceptable (and to whom) and can be realistically achieved? Moreover, is this kind of awareness-raising best done by scientists themselves or by other experts?

Some issues of praxis

We have spoken of how the internet expands the production of knowledge from science, how digital modes of exchange also offer new forms of community and new pathways to build and maintain it. Also increased is the contact between researchers and institutions important to their work, such as funding agencies, media outlets and also publishers. Yet none of this has come without struggle, especially regarding publications—the primary coin of intellectual capital for scientists.

During the late 1990s and early 2000s, in fact, the future of the scientific article became an issue of great conflict, as researchers and publishers battled over the legality of pre-print archives and open access. Many arguments by scientists, echoing early internet gurus, held that it is 'the nature of the internet' to make information open and free. Treating knowledge as 'intellectual property' (commercial material), it was said, is a rejection of core academic values. The call for open access has been backed, too, by the reality of the 'serials crisis'—runaway costs charged for scholarly journals, forcing libraries to abandon subscriptions and thus leave researchers (who supply journal content for free) in the lurch.

The conflict has made clear that any invocation of the internet as a final democratic essence is ingenuous: online 'space' is never free, but always paid for by someone, at some point, and is itself a form of investment. Scholarly societies have turned over most journal publication to commercial (for-profit) presses, because they could no longer accommodate the proliferation in specializations and research output. Outspoken scientists stress that the benefits of research come from access to findings and that society is worse off if such access is constrained. Publishers point out that such access has *always* been restricted in various ways, and that they are not charities. Science is thus a primary site where visions of the internet as an epistemological resource and commercial bounty come into deep conflict.

Debate continues about the role and limits of open access science. There are now a range of options in trial, including total open access and both partial and delayed open access. In many cases, journals that permit free admission do charge processing fees from authors or their institutions, thus in effect transferring the responsibility for such access (i.e. it is not free). Some allow authors to post a published article on their own web site or that of their institution, but with a link to the original journal site. A big question is whether any type of open access can be economically sustainable (opinions vary considerably). An alternative possibility, like the ISI Web of Knowledge, provides a single, subscription-based portal for access to the content of many different journals, plus other information like patents and conference proceedings. Other solutions are likely to be sought, as this remains an area of much concern (see Gartner, Chapter 3.2 this volume for further discussion).

Conclusion

The reality of internet science is now firmly established, rapidly evolving and full of possibilities and questions. There can be no doubt that the internet itself represents a new stage in human communication and that science benefits enormously, in fundamental ways, from it. Part of the reason is the quantity, variety and detail of the technical data, which are highly amenable to digital forms of recording and transfer. Moreover, the 1970s and 1980s saw the computer move to the centre of scientific work; thus when the internet fully arrived in the 1990s, science was ripe to not only embrace it but to explore and extend its capabilities.

Science is a hugely more public endeavour online, whatever the restrictions on access. This includes the general availability of scientific information to wider audiences and also the direct interaction between scientists and non-scientists in many domains—from personal opinion to formal debate over government policy. More open than ever before, science is also being asked via the online universe to be more engaged with other portions of society. Scientists have not recoiled from this expanded duty; on the contrary, they have sought to use the internet as a tool for public persuasion in ever more sophisticated ways, at times drawing on techniques from the mass media.

The future of scientific research, as well as science's global work and agency, is largely a digital, online future. Many important decisions still need to be worked out regarding the contours of this future. E-science remains contested territory in a number of areas. One must never ignore, either, the international dimension and all the unresolved matters there. In the end, online science is Big Science in a new idiom—more global than ever, more technologically dependent, yet because of these very facts, also more embedded in the immediate universe of politics, money and culture.

■ REFERENCES

Eisenstein, E. (1980). *The Printing Press as an Agent of Change*. Cambridge University Press, Cambridge.

Johns, A. (2000). *The Nature of the Book*. University of Chicago Press, Chicago.

Nye, D. (2006). *Technology Matters*. MIT Press, Cambridge, MA.

Pagels, E.H. (1992). *The Gnostic Paul: Gnostic Exegesis of the Pauline Letters*. Trinity Press, Philadelphia, PA.

Roberts, C.H. and Skeat, T.C. (1983). *The Birth of the Codex*. Oxford University Press, London.

■ FURTHER READING

• Borgman, C. (2007). *Scholarship in the Digital Age: Information, Infrastructure, and the Internet*. MIT Press, Cambridge, MA. Borgman has performed a considerable service in this book by discussing the internet as a form of social technology and tracing how it is used in all phases of a research project. She contrasts scholarly communication (continuous, many forms) with scholarly publishing (discontinuous, expanding number of forms) and talks about the essential role of infrastructure and the interrelationship of technological, social, institutional and intellectual demands. An excellent background work to any study of online science.

• Eisenstein, E. (1980). *The Printing Press as an Agent of Change*. Cambridge University Press, Cambridge. The indispensable work on the impact of printing in Western culture, Eisenstein's book covers all major aspects of this transformation: the technology and economics of early print, the important post-Gutenberg players, habits of reading, the role of authors and authorship, book culture in the Renaissance and literacy levels. Eisenstein has insights into every one of these topics, many of which are immediately relevant to the effects now being ascribed to the internet.

• Nye, D. (2006). *Technology Matters*. MIT Press, Cambridge, MA. Many of the issues surrounding the internet and science return to the role of technology in human communication. David Nye's book, simple and straightforward, provides an excellent lens through which to view such technology and what can be attributed to it. The book's early chapters on technological determinism are particularly helpful in critiquing the type of grand destiny so often misapplied to the internet generally.

• van Dijk, J. (2005). *The Deepening Divide: Inequality in the Information Society*. Sage, Thousand Oaks, CA. One of the world's leading experts on communication, van Dijk here lays out the issue of unequal access regarding the internet in today's world, including its general role in the sciences. The author shows that online inequality has many subtleties related to cultural and psychological factors—suggesting how this works within advanced nations too—as well as economic ones. Like Nye's book, this volume provides a sobering antidote to the 'world is flat' type of thinking. A suggestion arising from van Dikj's treatment is that different levels of support for online services among universities will create scientific inequalities too.

■ **USEFUL WEB SITES**

- *Nature* 'Access to the literature: the debate continues': http://www.nature.com/nature/focus/accessdebate/index.html. In 2004, the international journal *Nature* held a forum on the serials crisis in science and the issue of open access publication, mediated by Declan Butler. This site keeps online the various points of view expressed in that forum, by scientists, publishers, librarians and others. These brief essays and commentaries offer by far the best introduction and overview of this important topic, whose resolution is currently (2008) far from achieved today.

- **History of the Internet, Internet for historians (and just about everyone else):** http://www.let.leidenuniv.nl/history/ivh/frame_theorie.html. Despite its serious title, this is an eminently readable, well-organized narrative about the development of computers and the internet, and a bit more too. It begins with discussion of the nature of invention and its social influences, then moves on to early calculation machines, computers, ARPANET, e-mail, search engines and much more. Its main limitation is that it was written in 2002. For more updated sources see the list offered by the Internet Society at http://www.isoc.org/internet/history/.

- **The History of Visual Communication:** http://www.citrinitas.com/history_of_viscom/. This site offers nothing less than a visual history of human communication, from cave art to the computer. Divided into 10 separate chapters (Rocks and Caves, the Alphabet, Art of the Book, the Printing Press, etc.), its coverage is often brief but of high quality on a number of crucial subjects, and the imagery chosen as examples is not merely excellent but often mesmerizing. Science, art, technology and history join forces and offer many favours here.

- **World Summit on the Information Society (WSIS):** http://www.itu.int/wsis/index.html. In 2003 and 2005, two sets of important meetings on the global future of computer and internet technology took place, sponsored by the UN, with participants from many nations. The aim of this summit was to examine this technology as it has evolved in many parts of the world, and in many disciplines, and to create resolutions to help guide its continued global expansion. Among the specific topics discussed was the past, present and future of research and the role of the internet. (The specific URL for related presentations is http://worldsci.net/tunis/presentations.htm.)

3.2

From print to online: developments in access to scientific information

Richard Gartner

If I have seen further it is by standing on the shoulders of Giants.

Newton writing to Hooke (letter dated 5 February 1676)

No scientists, not even Newton or Hooke, have ever been able to divorce themselves fully from the work of their predecessors and contemporaries and research entirely in a vacuum; as a result, finding out what their peers have told the scientific world about their own work has always been part of the researchers' remit. How scientists find and use what might be called the corpus of scientific knowledge, the results and conclusions which others have drawn from their own research and made publicly available to the scientific community, is the subject of this chapter.

No scientist can be unaware that the most obvious change over the last 30 years in the way in which he or she interacts with the work of others is the move to the electronic provision of information in its many forms. This has altered scientists' interactions with the scholarly corpus in ways that extend beyond a mere change of medium: as well as the much larger volume of information now accessible, electronic media have brought new ways of searching and greater malleability to the information retrieved, offering great potential to empower the researcher. Whether their potential has been fully achieved, or whether constraints, personal or institutional, have prevented this, is a complex question.

In brief outline, this chapter starts with a look at the main media which researchers have traditionally used in print to access the scholarly record, and covers their extension from printed to electronic form. The rise of new movements that resulted from these technological developments, which aim to open up the provision of information and find new ways by which the academic record can be disseminated outside the traditional avenues of academic publishing, are then examined before a look at how the work of individual scientists has been affected by these changes. But first, it is necessary to take a close look at what the scholarly corpus is and how it is accessed.

The academic record in print

The corpus of academic achievement has traditionally been passed on through the print media of the academic monograph and the academic journal. Both have long been relied upon as the primary means by which research and its results are announced to the wider academic community and by which the research/results, and the scientists behind them, are assessed. Both have histories spanning centuries: 'scientific monographs' (in the broadest definition of the term) extend back to the works of the Greek pre-Socratic philosophers, whilst the scientific journal could be said to have started in England with the launch of the *Philosophical Transactions of the Royal Society* in 1665.

Despite their long history, these two media forms still predominate as the key ways in which research is recognized by the academic community. Their status as arbiters of value is usually seen as deriving from the role of peer review in their publication cycles: virtually all submissions to academic publishers undergo some degree of analysis by other academics in the same field, thereby providing what is considered to be an assurance of solid research value (see Wager, Chapter 4.1 this volume). The publication record of an academic is often crucial to career advancement, and an assessment of peer-reviewed publications is currently (2007–2008) a key part of the UK higher education community's Research Assessment Exercise (RAE).[1]

The size of the academic record enshrined in monographs and journals has generally meant that the scientist has had to approach its contents through some type of inter-mediary mechanism. Until the arrival of electronic publishing, this was usually through indexing and abstracting services, which digested the contents of academic publications and presented them in a more readily searchable form. Most academic journals have published subject indexes to the contents of their articles, either annually as supplements to their volumes (as has been the case with the journal *Nature* since it started in 1869) or in cumulative indexes covering a number of years (such as the 18th century index to the first 70 years of *Philosophical Transactions of the Royal Society* (Royal Society 1787). In addition, indexes and abstracting services which cover entire subjects, often includ-ing thousands of journals or books in their coverage, are also long-established ways of searching the academic record: *Chemical Abstracts*, for instance, a weekly abstracting service of the chemical literature, celebrated its centenary in 2007.

Before electronic information became so prevalent, the only way to use these services was through these printed index volumes. This was not an ideal state of affairs for scientists or for the librarians who had to provide them. Using an index entails trusting the indexer to understand the gist of an article accurately and then to provide appro-priate terms for it to be found, terms which often require the scientist to do a little homework to get to know and use properly. In the humanities, where opinions on what an article is 'about' and how it should be described can vary significantly from person to person, this can be a major problem: in the sciences, where terminology is generally more precise and the schemes of subject headings used to index works are generally more

---

1. http://www.rae.ac.uk/.

circumscribed than their equivalents in the humanities, these problems are generally less acute.

One other approach to searching the academic record is to use citation indexes such as the *Science Citation Index*. Instead of relying on third parties to analyse and index an article, these essentially list subsequent articles that cite it. The basic idea underlying this methodology is that these subsequent articles will tend to address the same subject and so will allow the researcher to carry out a literature search on a given topic by tracing the influence of any important articles on later research. This has the great advantage that it sometimes brings up articles which do not share any common words with the title or listed keywords of the original, and so allows greater recall of potentially relevant literature. It does, however, have the problem that citations do not necessarily indicate a strong congruence of ideas between articles, and in some cases, where a controversial work is cited principally in order to challenge it, the citation searches may have the opposite effect to that desired: for example the 697 current (2008) citations for Martin Fleischmann's (1989) article on 'cold fusion' are mainly negative.

Whatever the strengths and weaknesses of the philosophy behind these hardcopy indexing methods, their use in printed form proved an onerous business for substantial literature searches. The sheer bulk of the printed volumes of *Chemical Abstracts*, for instance, has long caused storage problems for librarians, while their users had the cumbersome job of searching multiple volumes (up to 26 volumes of weekly issues for recent articles, or 6-monthly volumes for older publications) to conduct their literature searches. Given the need, a particularly acute one in the sciences, to access material which is as current as possible, the time delays involved in indexing articles, and printing and distributing the indexes, also proved problematic.

The move to online

All of the problems associated with maintaining services of the type mentioned above inevitably led to pressures for delivery in the electronic rather than printed medium. Such a move became possible in the 1960s when electronic data storage as a means to the production of printed indexes became established practice: one of the first initiatives of this kind was MATICO (launched in 1963), a program to print library catalogue cards which was the work of Roger Summit, an early pioneer of online data bases (Bjorner and Ardito 2003).

Online access became publicly available in the early 1970s with the launch of two umbrella services, ORBIT (launched in 1971) and DIALOG (launched in 1972), which provided access to a small number of data bases including *Nuclear Science Abstracts* and *NASA RECON*. Both data bases used a complex command language which allowed Boolean searching,[2] a feature virtually impossible with printed indexes. Both services also charged for access by the second, and made further charges for each record viewed.

---

2. A method of searching which allows terms to be combined by operators such as 'and', 'or' or 'not' to allow greater precision and recall.

The charging structure for these services meant that in practice that they could only be performed by trained professionals, in most cases librarians who were taught online searching techniques as part of their library qualifications. Rather like surgeons performing amputations before anaesthetics, speed was of the essence, and the fast construction of searches and assessment of results became essential skills of the librarian. Nonetheless, an average search producing around 30 results would still result in charges in the region of $20–100, and so would not be undertaken lightly.

Online services of this kind were a great advance on the cumbersome processes of print indexes, but they were inevitably tightly restricted for cost reasons, often only to academics or doctoral students. It was not until the 1980s that technology in the form of the CD-ROM moved on to make end-user access to electronic data bases feasible for the first time. The CD-ROM offered what then seemed a large storage potential (approximately 650 MB, enough to contain most data bases on a single disk), and the possibility of designing more user-friendly interfaces which would allow the researcher to use them directly.

Unfortunately, allowing researchers direct access to CD-ROMs was far from straightforward. The first CD-ROM data bases were offered at high cost, generally much higher than their print equivalents, often forcing libraries to make decisions about whether to redirect funds from their book budgets to fund this new medium. Mostly CD-ROMs were published with single-use licences only, which meant that consulting them could be a laborious process involving checking out the disk like a book and loading it on to a dedicated machine: none of this was designed to make life as easy for the researcher as the technology allowed. Not for nothing did some sceptics as early as 1990 describe CD-ROM as a transient technology (McSean and Law 1990, p. 841).

Things improved slightly when CD-ROM networking technology arrived, allowing multiple access to the same data bases for the first time. For economic reasons, publishers were wary of opening access too freely in this way, and often imposed swingeing networking licence costs which had the effect of deterring many libraries from offering their users this enhanced service. The relatively primitive technology of CD-ROM networking also acted as an impediment to opening access, the slow retrieval speeds of the optical medium making them cumbersome to use for heavily used data bases, and the often flaky networking software causing its own problems of reliability. Certainly, as McSean and Law (1990) had pointed out, this technology had a built-in obsolescence.

The answer, it soon became apparent, lay in the development of data networking and its opening up to the academic and research communities. A number of small regional networks connecting academic institutions began to appear in the late 1970s in the UK, and were consolidated in 1984 into JANET (Joint Academic NETwork), which connected around 50 institutions together at what were then considered to be very high speeds. Although JANET was used mainly for e-mail and file transfers at that time, a few data bases, notably library catalogues, began to appear on the network, offering researchers real glimpses of the potential of online data bases for the first time.

In 1991, JANET became part of the burgeoning internet and interest in the online medium as a way of delivering data bases to the academic community really took off. Unfortunately, the terms on which the majority were made available remained as circumscribed as before, and few publishers felt the need to circumvent their profitable CD-ROM market by providing the same data online. It was only when the World Wide Web (the web) became well established in the mid-1990s, and graphical browsers such as Mosaic and Netscape began to make navigation and searching more user-friendly, that the delivery of end-user online data bases became something of a norm.

Often these services were provided by the same publishers as their CD-ROM predecessors, and began as alternative ways of delivering the same searching environment as the networked disks. Silverplatter's WebSpirs, for instance, which hosts such major data bases as *INSPEC* and *Biological Abstracts*, began as an additional way of accessing the company's extensive suite of CD-ROM publications. Today, the vast majority of data bases are delivered via the web, with CD-ROM very much the exception.

The academic record online

Online access to the academic record for individual researchers has undoubtedly profoundly changed their relationship to it. The academic journal and monograph, and the tools which index and abstract them, are now both potentially accessible electronically, with ramifications well beyond the mere change of medium. In particular, researchers and ancillary staff who support them (e.g. librarians) have been obliged to change their working practices to exploit the potential of these new media.

The electronic journal is probably the most immediately obvious manifestation of moves to access the academic record through online media. The vast majority of academic journals in the sciences are now available in online form, either as adjuncts to their printed versions or as solely electronic publications. Electronic provision does not, of course, equate to free provision, and most remain, as before, subscription peer-reviewed journals which continue to make major dents in libraries' budgets. Increasingly, academic books are being published in electronic form, most notably in the area of reference works, such as the *Encyclopedia of Astronomy and Astrophysics* (2007), which benefit the most from the facility of continuous updating that this medium allows.

The most obvious consequence of electronic provision of this kind is that accessing information no longer requires the physical proximity of the researcher to its source. This has had profound consequences for researchers and those, such as librarians, who traditionally made access to this information possible. For librarians in particular, the rationale of their very existence may be called into question if users can access information directly, and library budgets may come under attack if departments feel that they can subscribe to them directly rather than using the third-party services of a section outside their immediate control.

The most obvious functional advantage to the researcher in approaching the material electronically is the ability to use access points beyond those provided by an intermediary indexer. In using a printed index, the researcher is at the mercy of the indexer's choice of search terms; in its electronic equivalent any index point is potentially searchable, as is the full text of any abstract. The user may also be able to expand their searches, for example by using an integrated thesaurus to include synonyms (a facility available for *INSPEC* on WebSpirs, for instance) rather than relying on the indexer to provide references ('See also . . .') to find these.

In the case of electronic journals or books, the full text of the volume may itself be searchable. JSTOR, for instance, a subscription service which provides the contents of lengthy runs of over 1000 academic journals in electronic form, allows the user to search the full text of every page of every journal. This full text is also searchable under Google's search engines for those who subscribe to JSTOR, so allowing users to retrieve journal literature while conducting broader searches of the internet. Similarly, the growing number of academic e-books allow full-text searches of their entire contents.

The advantages of the opening up of new access points to the academic record cannot be overestimated, although it has brought some problems in its wake. Full-text searching in particular can be a notoriously inexact method of searching a corpus of writings. Even in the relatively exact terminological universe that the sciences inhabit, a full-text search will rarely be as precise as a search of descriptors assigned by an expert indexer, and will almost always involve extensive sifting through irrelevant results.

Despite these enormous changes in the provision of information to the scientist, the inherent model of what remains the canon of scholarly achievement has remained remarkably stable. The peer-reviewed journal is still seen as the essential medium for scholarly publication, and it is by their records as writers for these journals that academics, and the institutions they work for, are judged. The academic publishing industry remains as buoyant as ever, and the subscription-based journal remains the core of much of its business. The economic costs to libraries of providing access to this record have not lessened at all as a result of the move to electronic provision.

Such a continuation of previous models has not been to everyone's satisfaction. Many academic institutions, and the academics whose careers often depend on publication records, resent the huge costs involved in the traditional models of academic publishing and in some cases have made attempts to use the new technologies more imaginatively to circumvent some of these mechanisms and open up access. The form and success (or otherwise) of these 'open access' initiatives is the subject of the next section.

Towards open access

The provision of free access to electronic texts over networks can be said to stretch as far back as 1971 when Michael Hart created and networked his first e-text (Hart 1992), so starting *Project Gutenberg*, a library of electronic texts that continues to provide approximately 20,000 freely accessible electronic books to this day. Most of this is inevitably historical material which is out of copyright, although some authors, such as Brendan Kehoe (1992) whose *Zen and the Art of the Internet* was a key early work on the internet, have chosen it as their preferred medium of publication.

Moves to provide open access to contemporary academic output were pioneered in the early 1990s by academics who wished to make preprints of their forthcoming articles available for feedback and comment by their peers. The Los Alamos Physics Archive —arXiv—(Cornell University 2007) began in 1991 as a mechanism to allow researchers to self-archive preprints, although it now also contains archived versions of published papers, numbering over 446,000 as at July 2007. CERN (Organisation Européenne pour la

Recherche Nucléaire) in Geneva followed suit in 1993 and established an equally large archive to which all their researchers are required to submit preprints (Yeomans 2006).

Despite the work of such pioneering efforts, very few serious attempts to use the internet to open up access to the academic record were made during the 1990s, partly because of great reluctance on the part of publishers to cooperate with any initiatives for fear or losing revenue, and partly because of the traditional attachment on the part of academics to standard publication routes. Although some serial publications began to appear on the internet during this time, most of these tended to be newsletter-type publications, such as *Information Networking News*, an early journal about networking technologies in information science (JNT Information Networking Group 1992). It was not until the late 1990s, when freely accessible peer-reviewed journals such as the *JMIR* (*Journal of Medical Internet Research*, 2007) began to appear that the academic community made serious attempts to counter the overarching academic publishing model.

Various financial models have been tried to finance open-access journals while allowing them to resist the standard subscription method that most academic publications have traditionally followed. In many cases, the authors themselves or their funding bodies have to pay to submit their articles [the *JMIR*, for instance, charges a fee (at October 2007) of US$1500 for this]. Author-funded journals were found in a recent survey to make up 47% of open-access journals, the others receiving institutional funding or receiving revenue from advertising, membership fees or reprint charges (Suber 2006): one key example of this type is the Institute of Physics' *Journal of Physics: Conference Series* (Institute of Physics 2007), which uses the peer-review process of conference organizers to provide validation methods similar to those of traditional subscription-based journals.

Today, open-access journals are an established part of academic life, although they remain a relatively small proportion of electronic journals generally: for example, the *Directory of Open Access Journals* (*DOAJ*) (Lund University Libraries 2007), the most authoritative directory of scholarly journals of this type, lists only 39 journals in physics; by comparison, a 2008 search by this author of the Open University's library catalogue revealed a total of 727 subscription-based electronic journals in the same subject. Many of these open-access journals tend to be located outside Europe, including, in the case of physics, Asia, Iran, Russia, Brazil and India (Lund University Libraries 2007), where the traditional academic publishing model is not so entrenched.

Although open-access journals have not yet developed to a critical mass, institutional repositories which build on the pioneering work of Los Alamos and CERN have developed more significantly in recent years as academic institutions have begun to take the archiving and dissemination of their research output more seriously. One of the first initiatives of this type to gain noticeable attention was the DSpace project (released in 2002), a creation of the Massachusetts Institute of Technology (MIT) and Hewlett Packard (HP).[3] This project released an open-source software package which allowed institutions to set up and maintain their own repositories with a minimum of technical expertise. The project continues to this day, numbering 287 installations at October 2007,[4] and has

3. http://www.dspace.org/.

4. http://wiki.dspace.org/index.php/DspaceInstances.

been joined by other similar packages, including the SHERPA project's popular e-Prints software,[5] which is the most widely used system for repositories in the UK.

The easy availability of repository software and an increasing commitment by institutions to services of this kind has seen a rapid growth in their provision since 2002. By October 2007, 82 repositories were provided by the 131 higher education institutions in the UK, almost two-thirds of the total, and such is the momentum that the movement now has that most analysts expect 100% penetration of the sector within a few years.

Despite this increased institutional commitment to the idea of research repositories, take-up by academics remains patchy. At Oxford University, for instance, there were at the end of October 2007 only 742 items on the relatively new *Oxford Research Archive*,[6] less than one for every five members of the university's approximately 4000 research and academic staff.[7] Southampton University's *e-Prints Soton*,[8] one of the longest-established repositories in the UK, contained on the same date a much healthier 27,472 items, over 15 per member of the 1765 research and academic staff.[9] In part, this represents the success of a policy to require researchers to deposit a copy of any work to be used as part of the university's RAE assessment,[10] a policy not yet universally applied in the higher education sector.

If the actual volume of the academic record on open access remains relatively small, its importance is increasingly being recognized. The Budapest Open Access Initiative (Soros Foundation 2007), for instance, has been campaigning for strategic commitments to open access since its inception in 2002, and the prime international librarians' organization, the International Federation of Library Associations (IFLA 2004) has declared its support for the open-access cause.

Resistance from publishers does remain a significant impediment, with many imposing restrictions on the archiving of their publications in this way: Oxford University Press, for instance, currently (2008) impose a 12-month embargo on scientific articles before they can be made available in this way (Oxford University Press 2007). Vested interests may also manifest themselves among supposedly more disinterested bodies such as learned societies: one of the most dramatic examples in recent years is, perhaps, the Royal Society's (2005) negative statement on the subject which prompted a fierce rebuttal in an open letter from 64 notable scientists (Ashburner *et al.* 2005). The fact that such a groundswell of opinion can manifest itself certainly bodes well for the movement as it indicates strong levels of support amongst researchers themselves who resent many features of the traditional model, although institutional resistance, or at least inertia, will undoubtedly be a major impediment for years to come.

5. http://www.sherpa.ac.uk/.

6. http://ora.ouls.ox.ac.uk/access/.

7. http://www.ox.ac.uk/about_the_university/facts_and_figures/index.html.

8. http://eprints.soton.ac.uk/notes.html.

9. http://users.ecs.soton.ac.uk/hnr/UCASpresentastion.ppt.

10. http://eprints.soton.ac.uk/.

Given the vested interests stacked up against it, what are the chances of the principles of open-access initiatives changing the researcher's relationship to the academic record to any substantive degree? Certainly for as long as peer-review academic publication remains the dominant means of assessing the value of research, some method of finance will be required for what is an expensive and still cumbersome process, and open-access journals will continue to require funding either by contributors or institutions. Institutional repositories offer the possibility of much cheaper access to the academic record, but not all material on them comes with what is perceived as the essential quality assurance of this method of validation.

Could there perhaps be alternative means of validating academic worth without peer review which would make use of the full potential of these new technologies? This is a question whose examination in depth is beyond the scope of this chapter, but the technology undoubtedly exists for the academic community to assess scholarly works using review methods which are already common currency in such services as Wikipedia,[11] or even online shops such as Amazon.[12] Making full use of these will depend primarily on the willingness of academics to take the time to approach the literature in a more investigative way, which in itself will require increasingly sophisticated skills in information retrieval and analysis on their part. It will also require an increased willingness on the part of academic employers to assess the value of research without over-reliance on lists of peer-reviewed publications. Changes in scholarly culture of this kind can only come with time, and will inevitably lag behind technological advances.

The researcher's perspective

Assessing the impact of these changes on the researcher's relationship to the academic record can be done relatively easily in a quantitative sense: a simple examination of usage statistics, as has been done in many studies (such as Mele *et al.*'s (2006) survey of the publishing landscape in high-energy physics) is enough to show how pervasive the electronic medium is as the prime conduit for communicating academic discourse. Similarly, a reasonable picture of the success (or otherwise) of the open-access movement can be gleaned from number-crunching of the type outlined above.

The qualitative experience of these new media is harder to gauge, and often may only be assessed by documenting the experiences of individual scientists. A small number were interviewed for this chapter about their methods and approaches to querying the corpus of academic knowledge by electronic means in an attempt to gain a sense of how it has changed their relationship to the work of their peers. As might be expected, these scientists work almost exclusively with electronic information sources, with only a few print journals, notably *New Scientist* (a non-peer-reviewed journal which does not publish original research), retaining widespread readership. None, however, read any

11. http://www.wikipedia.org/.
12. http://www.amazon.co.uk/.

open-access journals on a regular basis, and none had submitted articles to one, although almost all felt positive about their potential. Of course, this is not intended to be a comprehensive survey of current practices in accessing information, merely indicative of the challenges involved in embedding new practices.

The usage of institutional repositories was somewhat higher than that of open-access journals, although still patchy at best. One Open University lecturer made the most extensive use of these, including the university's own[13] and those of several US universities, but only a 'few times a year'. Others preferred to use specialized services, such as *SCIRUS* (Pruvost *et al.* 2007), which search digital resources across multiple sites rather than accessing repositories directly. Few had submitted items to their institutional repository, although they were encouraged to do so by their host institutions: often such items as were submitted were joint papers, sometimes with as many as 600 authors.

The findings of this small survey appear to accord with those of more extensive research into scientists' relationships with open-access publishing. A major study for the UK's Joint Information and Systems Committee (JISC), for instance, found a similar mix of reasons for authorial resistance to open-access journals, including a perception (not borne out by reality as the study shows) that readership and citation levels for these journals are necessarily lower than those for more traditional ones (Swan and Brown 2004). The most common reason, however, for authors' limited engagement with open-access publishing was found to be a lack of awareness of the existence of open-access journals in their field of expertise, a situation exacerbated by the rather haphazard ways (often relying on the recommendations of colleagues above all else) in which they tended to find out about them (Swan and Brown 2004).

The question of publication fees for open-access publications remains a contentious one. Although Swan and Brown (2004) found that the majority of authors did not have to pay a fee themselves (by virtue of their institutions being members of the organization publishing the journal, for instance) and did not consider fees an impediment to their submitting to these journals, there is genuine concern that the fee-based model may make it difficult for young researchers without extensive grants or for those from developing countries to publish. Although many journals have policies of waiving or reducing fees for authors from developing countries, this may not prove a panacea: a study of the views of academics at the University of the West Indies towards open-access journals, for instance, found that many felt embarrassed to claim the waiver or that when they did, their submissions were often delayed or given lower priority (Papin-Ramcharan and Dawe 2006). The study also found that waiver policies are rare in narrower disciplines where the range of relevant journals is limited. All of these factors produce impediments for authors in the developing world to compound those already caused by limited technical infrastructure and other barriers to access, but they should by no means be insuperable.

For the individual researcher, there may be many other reasons beyond those given above for a reluctance to publish via the open-access route. Partly this may be a reflection of the attitude of the researcher's institution, which obviously has a direct bearing

13. *Open Research Online*, http://oro.open.ac.uk/.

on its staff's careers: if, for example, tenure committees give less weight to publications in open-access journals, few academics would risk their future career prospects by submitting to them. Such attitudes are, however, undoubtedly in retreat, with fewer and fewer academics seriously considering a peer-reviewed open-access journal less worthy than its more traditional counterpart.

More practical problems also play a part for the individual scientist. Information overload appears to be a problem for many, a situation not helped by the number of sites with potential relevance to any subject and the fact that most use different interfaces with which the researcher has to become familiar in some depth before they can be used to their full capability. These problems can be addressed from various angles, by, for instance, investment in generic cross-searching services such as *SCIRUS* mentioned above, which needs to be taken up by the information professions, and by the scientists themselves in recognizing their own needs for new skills in finding and using information and taking the initiative to acquire them. Above all, the open-access movement needs to work continuously on consciousness raising, without which it will inevitably founder: scientists and institutions will not make the commitments outlined above until it is firmly established as an equal, if not primary, medium of communication.

Conclusion

The overall picture that could be seen to emerge from this history and these surveys, both quantitative and qualitative, is of a potential not realized. The open-access movement has impinged only minimally on the dominance of traditional publishing channels for academic research, and institutional repositories are still at a relatively early stage of development, although their impact is undoubtedly being felt throughout the academic sector. The empowering effects of online access to the scholarly record, including more widespread availability and new ways of finding information (such as full-text searching) that circumvent many of the restrictions imposed by the print medium, cannot be disputed; despite this, the over-abundance of information that has come in the wake of these technological developments could be seen as a debilitating counterpart to the researcher's new powers.

This need not be seen as an overly pessimistic picture: technological advances often precede the changes in attitude, personal or institutional, which are needed to make full use of their potential. The statement by the Budapest Open Access Initiative, that 'an old tradition and a new technology have converged to make possible an unprecedented public good' (Soros Foundation 2007), is not entirely wishful thinking, although its intentions may not be realized for some while yet. The open-access movement has yet to flex its muscles in some key areas, particularly in relationships with academic publishers: while the subscription journal or traditionally published academic monograph remains the prime method of validating academic worth, open access will always be handicapped to some degree, although, as has been shown with journals such as the *JMIR*, the open-access model is not incompatible with the traditional method of peer review.

What are the prospects for technology realizing its possibilities in opening up information in the foreseeable future? Certainly the online repository is becoming an important adjunct to traditional academic publishing mechanisms and is certain to become increasingly so now that it is receiving institutional support at senior levels. The prospect for open-access journals looks less rosy: their limited impact so far, and the likelihood that most journal articles will appear, at least after an embargo period, on repositories may call their *raison d'être* into question.

Ultimately the form of media used by the open-access movement is less important than the fact of its openness: however the scholarly record is opened up, the beneficial effect, including the mitigation of the financial restrictions that previously skewed the benefits of scholarly research to those institutions or societies that could afford to subscribe to traditional media, will no longer predominate. Perhaps the overall vision of the Budapest initiative, to 'unit[e] humanity in a common intellectual conversation and quest for knowledge' (Soros Foundation 2007), is already being achieved.

■ REFERENCES

Ashburner, M. *et al.* (2005). *Open Letter from Fellows of the Royal Society to its President* [online]. Available at: **http://www.frsopenletter.org/**.

Bjorner, S. and Ardito, S.C. (eds) (2003). Online before the Internet: early pioneers tell their story. *Searcher*, June [online]. Available at: **http://www.dialog.com/about/history/pioneers1.pdf**.

Cornell University (2007, November 16). *Lanl.arXiv.Org e-print Archive Mirror*. Available at: **http://xxx.lanl.gov/**.

Encyclopedia of Astronomy and Astrophysics (2007). *Encyclopedia of Astronomy and Astrophysics*. Taylor and Francis, London (see: **http://eaa.crcpress.com/**).

Fleischmann, M. (1989). Electrochemically induced nuclear fusion of deuterium. *Journal of Electroanalytical Chemistry*, **261**, 301–8.

Hart, M. (1992). *Gutenberg: the History and Philosophy of Project Gutenberg*. Available at: **http://www.gutenberg.org/wiki/Gutenberg:The_History_and_Philosophy_of_Project_Gutenberg_by_Michael_Hart**.

IFLA (International Federation of Library Associations) (2004). *IFLA Statement on Open Access to Scholarly Literature and Research Documentation*, 7 December. Available at: **http://www.ifla.org/V/cdoc/open-access04.html**.

Institute of Physics (2007). *Journal of Physics Conference Series*. Available at: **http://www.iop.org/EJ/conf**.

JNT Information Networking Group (1992). *Information Networking News*, Issue 1 (May). Available at: **http://www.ibiblio.org/pub/docs/about-the-net/libsoft/infnetnews1.txt**.

Journal of Medical Internet Research (2007). **http://www.jmir.org/**.

Kehoe, B.P. (1992). *Zen and the Art of the Internet*. Project Gutenberg, Lisle.

Lund University Libraries (2007). *Directory of Open Access Journals*. Available at: **http://www.doaj.org/**.

McSean, T. and Law, D. (1990). Is CD-ROM a transient technology? *Library Association Record*, 92(11), 837–8, 841.

Mele, S., Dallman, D., Vigen J. and Yeomans, J. (2006). Quantitative analysis of the publishing landscape in high-energy physics. *Journal of High Energy Physics*, **12** [online]. Available at: **http://doc.cern.ch//archive/electronic/other/generic/public/cer-002661252.pdf**.

Oxford University Press (2007). *Oxford Journals Access and Purchase: Author Self-Archiving Policy*. Available at: **http://www.oxfordjournals.org/access_purchase/self-archiving_policya.html**.

Papin-Ramcharan, J.I. and Dawe, R.A. (2006). Open access publishing: a developing country view. *First Monday*, **11**(6) [online]. Available at: **http://firstmonday.org/issues/issue11_6/papin/index.html**.

Pruvost, C., Knibbs, E. and Hawkes, R. (2007). *SCIRUS—for Scientific Information Only*. Available at: **http://www.scirus.com**.

Royal Society (1787). *A General Index to the Philosophical Transactions, Vol. 1–70*. Royal Society, London.

Royal Society (2005). *Royal Society Position Statement on 'Open Access'*. Available at: **http://royalsociety.org/page.asp?id=3882**.

Soros Foundation (25 June 2007). *Budapest Open Access Initiative*. Available at: **http://www.soros.org/openaccess/read.shtml**.

Suber, P. (ed.) (2006). No-fee open access journals. *SPARC Newsletter*, **103** [online]. Available at: **http://www.earlham.edu/~peters/fos/newsletter/11-02-06.htm#nofee**.

Swan, A. and Brown, S.N. (2004). *JISC/OSI Journal Authors Survey Report*. [Online]. Available at: **http://www.jisc.ac.uk/uploaded_documents/JISCOAreport1.pdf**.

Yeomans, J. (ed.) (2006). CERN's open access e-print coverage in 2006: three quarters full and counting. *High Energy Physics Libraries Webzine*, Issue 12 [online]. Available at: **http://library.cern.ch/HEPLW/12/papers/2/**.

■ FURTHER READING

- Gorman, G.E. (ed.) (2005). *Scholarly Publishing in an Electronic Era*. Facet, London. A well-chosen collection of articles on current trends in academic publishing, including a section on the open-access movement.

- Jacobs, N. (ed.) (2006). *Open Access: Key Strategic, Technical and Economic Aspects*. Chandos, Oxford. A collection of articles by some key figures in the field, this volume is very strong on the economics of open access. In the spirit of the movement, all the chapters have been self-archived in institutional repositories.

- Jones, R., Andrew, T. and MacColl, J. (2006). *The Institutional Repository*. Chandos, Oxford. A book aimed at information practitioners, this offers pragmatic advice on the issues underlying institutional repositories based on the experiences of those who have already established them.

- Swan, A. and Brown, S.N. (2004). Authors and open access publishing. *Learned Publishing*, **17**(3), 219–24. A review of authors' attitudes to open-access publishing, based on a number of surveys, revealing disparities of awareness of 'self-archiving' in particular.

- Willinsky, J. (2006). *The Access Principle: the Case for Open Access to Research and Scholarship*. MIT Press, Cambridge, MA. A survey of issues surrounding the open-access movement that covers broader issues, including human rights and the principle of public access to knowledge in general.

■ USEFUL WEB SITES

- arXiv: **http://arxiv.org/**. An example of an online preprint archive, arXiv is one of the longest-established and most comprehensive in its subject areas (physics, mathematics, computer science, quantitative biology and statistics).

- **Directory of Open-access Journals:** **http://www.doaj.org/**. The most comprehensive list of open-access journals currently available: approximately one-third of its titles are searchable at article level.

- *Open*DOAR: **http://www.opendoar.org/**. A selective directory of open-access institutional repositories of relevance to academic research, a sister project to the Directory of Open-access Journals.

- **ROAR, Registry of Open-access Repositories:** **http://roar.eprints.org/**. The most comprehensive international directory of institutional repositories, including the ROAR search engine, which provides a Google-type search of many of their contents.

- **SHERPA:** **http://www.sherpa.ac.uk/**. A partnership of 26 institutions to develop and promote institutional repositories within the UK higher education sector: its services include *Open*DOAR (listed above) and SHERPA Search, a simple full-text search service for UK repositories.

SECTION 4

Consensus and controversy

The problem with experiments is that they tell you nothing unless they are competently done, but in controversial science no-one can agree on a criterion of competence. Thus, in controversies, it is invariably the case that scientists not only disagree about results, but also about the quality of each other's work.

Harry Collins and Trevor Pinch (1993) *The Golem; What Everyone Should Know About Science*

4.1 **Peer review in science journals: past, present and future,**
     by Elizabeth Wager   115

4.2 **Controversy and consensus,** *by Jeff Thomas*   131

4.1

Peer review in science journals: past, present and future

Elizabeth Wager

Introduction

Peer review can be traced back almost 300 years and is used throughout academia to decide which articles appear in scholarly journals and which results get presented at learned conferences. It is also used by research funders to determine which projects they should support and by publishers to decide which books to publish. In most disciplines, the output and productivity of individual academics and university departments is measured in terms of the number of papers they have published in peer-reviewed journals or presented at conferences. Peer review therefore not only determines the shape of the research record (and, therefore, the direction of future research), but can also have a major influence on academic careers. While the systems of peer review used by different disciplines and for different purposes are similar, this chapter focuses on peer review in science journals and, in particular, on biomedical publications, since the majority of research into peer review has focused on this area.

What is peer review?

When an article is submitted to a journal, the editor contacts one or more reviewers (or referees) considered to be experts in their field and asks them to comment on the article. The term 'peer' denotes the author's equals. The World Association of Medical Editors (WAME) noted in 2007 that 'A peer-reviewed biomedical journal is one that regularly obtains advice on individual manuscripts from reviewers who are not part of the journal's editorial staff'.[1] Reviewers are therefore expected to be independent, both of the journal and of the work under scrutiny.

1. http://www.wame.org/resources/policies#definition.

Reviewers are generally asked whether the work is methodologically sound, original and clearly reported. Reviewers may also be asked to comment on the importance of the research and whether the findings are likely to interest the journal's readers. However, in most journals, the final decision about whether to accept (and publish) or to reject a submission is made not by the reviewers but by the editor or an editorial committee. For this reason, the terms 'reviewer' and 'peer review' tend to be preferred over 'referee' and 'refereeing', since, in other arenas, referees act as arbiters and decide whether an action is acceptable or not, whereas this is not usually the role of peer reviewers who generally act only as advisors to the editor.

Most journals do not pay their peer reviewers since researchers are generally prepared to review other people's work on the understanding that their own submissions will receive similar treatment. Highlighting the reciprocity of the system, and the difficulty of finding suitable reviewers, some editors have suggested that they should refuse to consider submissions from researchers who repeatedly refuse to review other people's work (*MEPS* 2000).

For a journal to be considered peer reviewed, all published reports of scientific research must have been scrutinized by suitably qualified, independent scientists. However, since reviewers are usually not paid, and it may be difficult to find enough suitably qualified researchers prepared to review an article, some journals reserve external review for the submissions they are most likely to publish and reject others on the basis of the editor's opinion alone.

This system of preliminary in-house review is used mainly by journals with full-time editorial staff and those that receive large numbers of submissions. For example, *The Lancet* rejects 80% (according to a personal communication from the journal's Senior Executive Editor, Sabine Kleinert) and the *Journal of the American Medical Association* (*JAMA*) rejects about 60% of submissions on the basis of in-house review (Fontanarosa and DeAngelis 2007). While full-time journal editors are likely to know what will interest their readers, they cannot be experts on every paper they receive, so one could argue that, in journals that use this method, while acceptance is based on review by one's peers, rejection might not be.

Even when submissions are critiqued by external reviewers, the definition of a peer or equal is a subjective one. Potential reviewers are usually selected on the basis of their publication record and reputation. Editors virtually never check the reviewer's qualifications and, in many cases, may not know the reviewer's professional status or position (although they will usually know their academic affiliation). It is therefore not possible to guarantee that professors' work is reviewed only by other professors, and so on. Thus, the definition of a peer is not a precise one.

Interestingly, while researchers may wish to have their work reviewed by people they consider their equals or superiors, there is evidence that the most eminent scientists are not necessarily the best reviewers. Black *et al.* (1998) studied reviews for a medical journal and found that the best reviews (according to the journal's editors) were generally produced by younger scientists (aged under 40) and those with formal training in epidemiology or statistics, rather than by those holding the most senior positions.

One problem in identifying appropriate reviewers is that researchers working in closely related, highly specialized fields may be unsuitable since they may be competing with the person or institution whose work they are expected to review impartially. The authors of a chapter about the peer review of grant applications concluded 'there is no such thing as the perfect reviewer. Those too close to the subject may be influenced by jealousy or cronyism. More distant, and they may suffer from lack of expertise' (Wood and Wessely 2003, p. 14). Similarly, the editor of the *British Medical Journal* (*BMJ*) has written 'I have heard it argued that the only person who does not have some sort of vested interest in a subject is somebody who knows nothing about it at all' (Smith 1994, p. 5).

Another potential problem of being reviewed by one's peers is that such review may fail to recognize exceptional or unconventional work (see also Lewenstein, Chapter 5.1 this volume for discussion). Since reviewers are generally selected for their success and expertise in traditional methods and accepted theories, they may fail to appreciate, or even feel threatened by, radical advances and new hypotheses. The computer scientist Edsger W. Dijkstra commented in 1987 that 'Not only does the mechanism of peer review fail to protect us from disasters, in a certain way it guarantees mediocrity: the genius has no peers'.[2]

This problem is not a new one. A paper submitted to the Royal Society by the relatively unknown scientist J.J. Waterston in 1845 described ideas that later formed the basis for the kinetic theory of gases published by Joule, Clausius and Clerk Maxwell. Yet the reviewer of Waterstone's ground-breaking manuscript commented 'this paper is nothing but nonsense' so it was never published (Lock 1991, p. xxiii).

More recently (in 2003), an editorial in *Nature* noted that this prestigious science journal had 'a confession to make' relating to a publication by Paul Lauterbur who had just won the Nobel prize for medicine. The journal 'pleased with having published Lauterbur's work . . . celebrated it along with other *Nature* greats in a promotional campaign. Lauterbur politely wrote in to point out that we had published only after he had appealed against a rejection' (Anon 2003, p. 645). In case you think this might be an isolated incident, there is even a web site chronicling articles from Nobel prize winners that have been rejected by journals.[3]

The choice of reviewers is certainly subjective, yet it lies at the heart of the peer-review system. Selecting reviewers is probably one of the most important skills of an editor, yet very few editors receive any training in this. Some journals ask authors to suggest potential reviewers and this may be a good strategy, since authors may be better at identifying their peers than editors. Although editors may be suspicious that authors will suggest reviewers likely to give them an easy ride, studies have shown that the reviews produced by author-nominated reviewers are of similar quality to those from reviewers chosen by editors (Wager *et al.* 2006).

2. http://www.cs.utexas.edu/users/EWD/transcriptions/EWD10xx/EWD1018.html.

3. http://www2.uah.es/jmc/nobel.html.

When did journals start to use peer review?

The records of the oldest European scientific societies show that systems of peer review almost identical to those used today were in place as early as the late 17th century (Kronick 1990). By the first half of the 18th century many learned societies employed review by experts to determine which reports would be presented at the society's meetings and published in their journal (often called proceedings or transactions). At that period, science was classed as a branch of 'natural philosophy' so these journals often have titles such as 'Philosophical Transactions' yet contain accounts of scientific experiments.

It is not possible to give an exact date for the start of peer review, but by 1731 the Royal Society of Edinburgh announced that papers would be 'distributed according to the subject matter to those members who are most versed in these matters. . . . Nothing is printed in this review which is not stamped with the mark of utility'. Similarly, in 1752, the Royal Society of London established a Committee on Papers which reviewed submissions for the society's journal *Philosophical Transactions* (which had been started in 1665 and probably used some sort of peer review before 1752). The committee (which required five members for a quorum) could also call on 'other members of the Society who are knowing and well skilled in that particular branch of Science that shall happen to be the subject matter of any paper which shall be then to come under deliberations' (quoted in Kronick 1990, p. 1321). Across the Channel in France, by 1782, the Académie Royale de Médecine had also set out details about how submissions were reviewed before publication (Kronick 1990).

Although it is clear that a recognizable form of peer review was well established in many European learned societies by the 17th or 18th centuries, right up to the middle of the 20th century many science and, in particular medical, journals did not routinely apply external review to all published articles (Burnham 1990). Until then, individual editors often accepted articles without the benefit of an external opinion or depended on a small in-house staff for most editorial decisions. For example, in the 1940s the editor of the *JAMA* noted that the journal sought expert opinion 'rarely indeed' (Burnham 1990, p. 1327). Whether or not a journal routinely used external review seems to have depended, to a large extent, on the personal preferences of the editor. In 1893, Ernest Hart, the editor of the *BMJ*, addressing a meeting of American medical editors stated that every article in his journal was 'referred to an expert having special knowledge and being a recognized authority in the matter'. He noted that this was 'a laborious, difficult method, involving heavy daily correspondence and constant vigilance to guard against personal eccentricity or prejudice'. However, he considered these efforts worthwhile to ensure 'authoritative accuracy, reality and trustworthiness' (quoted in Burnham 1990, p. 1325).

Although Hart considered that peer review could ensure the trustworthiness of publications, it is interesting to note that, even in the 18th century, editors recognized that peer review was ineffective in detecting fraudulent research. The Literary and Philosophical Society of Manchester, in establishing its Committee of Papers in 1785 noted that review could guarantee only 'the novelty, ingenuity, or importance' of the submissions and that 'Responsibility concerning the truth of facts, the soundness of reasoning, . . . [and] the accuracy of calculations is wholly disclaimed: and must rest alone, on the knowledge, judgement, or ability of the authors who have respectfully furnished such communications' (Kronick 1990, p. 1322).

How is peer review performed in science journals today?

Although peer review can be broadly defined as critique of submissions by external experts, the actual procedures employed by journals vary widely. The outcomes are also very varied, with acceptance rates ranging from 5% to 85%. This is another reason why it has proved almost impossible to measure the quality of peer review since there is no universally agreed standard by which to measure how well it is performing.

The Association of Learned and Professional Society Publishers (ALPSP) together with the European Association of Science Editors (EASE) surveyed their members in 2000 and the 200 responses demonstrated the wide variations in how peer review is practised across science journals (ALPSP/EASE 2001). Most journals (73%) used two reviewers for each article but 6% used only one, 18% used three, and six journals (3%) used more than three. In 53% of journals one of the editors sometimes or always acted as a referee. The final decision on acceptance or rejection was made by the editor in 84% of the journals and by the editorial board in 13%. Acceptance rates ranged from under 10% to over 75%.

Another variation between journals is whether peer review is anonymous. It is interesting to note that in 1731 the Royal Society of Edinburgh noted that 'The identity [of the peer reviewers] . . . is not known to the author' (Kronick 1990, p. 1321) and it has traditionally been the case that reviewers' identities are not revealed to authors. The ALPSP survey showed that 88% of journals continued to use this method of review in 2000.

Another variation is whether authors' names are left on the paper or are removed before it is sent to reviewers. Some journals (40% in the ALPSP survey) use what may be called masked or blinded review (i.e. the authors' names are not revealed to the reviewers) in an attempt to reduce reviewer bias, since there is evidence that reviewers may be biased against studies coming from smaller institutions or from overseas countries. There may also be bias based on the authors' gender or qualifications (see the section on 'Is peer review fair?' for more details).

One practical problem with masked review is that it may be possible to guess the authors' identity even though their details have been removed from a paper. This is because researchers tend to cite their previous work, so the authors' names often appear in the reference list and, in small specialities, researchers know each other and have an idea about what other people are working on.

Reviewing all the studies about the evidence for and against bias and the effects of removing author names from medical papers, Robert and Suzanne Fletcher (former editors of the *Annals of Internal Medicine*) concluded 'journal editors might reasonably choose to blind or not. There appears to be little at stake in their choice' (Fletcher and Fletcher 2003, p. 69).

Does peer review work?

The primary purpose of peer review is to help journal editors select articles for publication or for funders to decide which projects to fund. An obvious analogy for such selection processes is medical diagnostic tests. The ideal test for any condition (e.g. AIDS or prostate cancer) will always pick up the disease if it is really there (i.e. there will be no false negatives) and will never give a false-positive result suggesting you have the disease when, in fact, you do not. In other words, the test should have high sensitivity and specificity. In terms of science journals, editors want a test capable of detecting all high-quality papers and of rejecting all seriously flawed ones. Editors also want a reliable test, i.e. one that will give the same answer if repeated several times. One way of testing whether peer review works is therefore to measure how often reviewers agree with each other.

Several studies (mainly in biomedical journals) have shown that peer review, measured in this way, is not particularly reliable. In other words, reviewers often disagree about the merits of a submission. Given this variability it has been calculated that, to obtain a statistically reliable decision, editors would need six reviewers, all agreeing on whether to accept or reject an article (Fletcher and Fletcher 2003). Yet, in practice, most editors obtain only two or three reviews for each submission and reviewers often disagree in their recommendations.

Publishing decisions therefore rely to a considerable extent on the individual journal editor or editorial board, and the views of peer reviewers may be ignored or overruled. Many respected journals reject more than 90% of the articles they receive; editors admit that this means they may miss important work and that sound submissions are often rejected simply because the findings are not considered sufficiently exciting or interesting to readers. Peer review, as performed by journals, certainly cannot be considered an exact science.

Yet even with medical tests it is rare to find one with 100% sensitivity and specificity, so perhaps it would be better to focus on eliminating grossly misleading findings. In this case we might measure success in terms of whether peer review is effective at identifying fraudulent research. Once again, the evidence is not convincing. Papers containing fabricated data have been published in the top science journals. For example, the discredited Korean cloning studies of Hwang Woo-Suk and the fraudulent nanotechnology experiments of Jan Hendrik Schön were both published in *Science* (the highly respected peer-reviewed journal of the American Association for the Advancement of Science). An investigation into the work of Norwegian dentist John Sudbø found that he had published fabricated data in both *The Lancet* and *The New England Journal of Medicine* (two of the most highly rated medical journals) (Nylenna and Horton 2006).

So, it seems that peer review is not particularly effective at stopping fraudulent science from getting published. The publication of fraudulent research in biomedicine is of particular concern since clinical decisions and guidelines may be based on the results of single studies. In particular, because they are expensive and time-consuming to conduct, large-scale clinical trials are often never repeated so their findings are rarely tested in this way. However, in other disciplines, such as the physical sciences, single rogue papers may not have such long-term importance because advances are accepted only after the initial experiments have been replicated in different laboratories, so the literature is 'self-correcting'.

While many editors recognize that peer review is relatively poor in spotting fraudulent data (which, so far as we can tell, is relatively rare), they argue that it has a value in raising the standards of reporting for the majority of valid studies. But another study casts doubt on this. Editors from the *BMJ* inserted eight deliberate (and quite serious) errors into a research paper and sent it to over 200 of the journal's regular reviewers who were unaware that they were taking part in an experiment. On average, the reviewers spotted only two of the eight errors, just 10% spotted at least half of them, and 16% did not identify any at all, which is hardly reassuring (Godlee *et al.* 1998). However, it is possible that reviewers perform better in 'real life', since one of the difficulties of this type of experiment is identifying enough reviewers with sufficient expertise to review the same paper. It is possible that the reviewers in the *BMJ* experiment failed to spot weaknesses in the paper because they were not sufficiently familiar with the topic. But, in that case, since the reviewers were unaware they were taking part in an experiment, one might have hoped that, being aware of their limitations, they would simply have refused to review the paper.

In an attempt to summarize all the available evidence and discover whether peer review in medical journals is effective, Jefferson *et al.* (2002) assessed all published studies about the effects of peer review on the quality of reporting medical research and concluded that peer review had 'uncertain outcomes'. Discussing the study in *The Guardian*, Jefferson went on to say that 'If peer review were a medicine it would never get a licence. . . . We had great difficulty in finding any real hard evidence of the system's effectiveness, which is disappointing, as peer review is the cornerstone of editorial policies worldwide' (quoted in Petit-Zeman 2003).

Is peer review fair?

Given the importance of peer review in publishing decisions and academic careers, it would be reassuring to know that the system, even if it may be imperfect, is at least objective and unbiased in terms of authors' gender, age, origin and so forth.

One of the most striking studies suggesting the existence of reviewer bias was performed by Peters and Ceci (1982). They took 12 articles that had already been published in psychology journals by authors from prestigious American institutions. They retyped the papers and resubmitted them to the journals that had already published them after changing the authors' names and affiliations to less impressive-sounding (fictitious) institutions. Only three journals noticed that they had published the paper before. Of the other nine papers, eight were rejected, but on grounds of weak study design or inappropriate statistical analysis rather than lack of originality, leading Peters and Ceci (1982) to conclude that the reviewers were biased against authors from unknown institutions.

However, other studies have shown that removing authors' details has no effect on the quality of the review and that biases may work in the opposite direction to that suggested by Peters and Ceci. For example, Fisher *et al.* (1994) found that well-known authors received harsher reviews than those who had not published so much.

There have been surprisingly few studies looking at gender bias, so it is hard to generalize, but the findings of those few are far from reassuring. Margaret Lloyd (1990) investigated the effect of gender on decisions by 65 reviewers for behaviour journals by sending out papers on which the authors names had been changed (so all reviewers received the same paper except that in half the cases it was apparently written by two women and in the other half by two men). She found that women reviewers recommended acceptance of the paper more often if the author were female (when 62% recommended acceptance) than if the authors were male (when only 10% recommended acceptance). The male reviewers did not respond differently to the male- and female- authored papers, but only 21% of them recommended acceptance of either version. This study therefore suggested that female reviewers tended to be biased in favour of women, or against men, or perhaps both.

Wennerås and Wold (1997) investigated the Swedish Medical Research Council's system for awarding grants and concluded that the system was biased against women. They found that female applicants got lower scores, on average, for scientific competence, than male applicants, but these scores were not justified by the applicants' publication record. They concluded that women had to be 2.5 times more productive than men (in terms of their publication output) to achieve the same score.

There have been other studies of student perceptions of papers, which suggest that women are biased against papers by other women (i.e. the opposite of what Lloyd (1990) found) but few others looking directly at gender bias by peer reviewers. However, Kay Dickersin and colleagues (1998) showed that the proportion of women on the editorial boards of four American epidemiology journals (13%) was lower than the proportion of women who published in the journal as authors (29%).

While rejection by a journal is disheartening for the authors, there are usually plenty of others to choose from, so persistent researchers tend to get their work published eventually. However, applying for research funding is often extremely time-consuming, and each funder has their own set of complex forms and requirements, so arbitrary or unfair funding decisions can have more serious consequences. Despite the studies mentioned, a review of the evidence about the peer review of grant applications concluded that 'peer review processes as operated by the major funding bodies are generally fair' (Wood and Wessely 2003, p. 14). However, an overview of bias in journal peer review (published in the same book) reached the opposite conclusion, suggesting 'despite the paucity of direct evidence [to bias in editorial decisions] . . . the balance of evidence suggests that many of these biases do exist' (Godlee and Dickersin 2003, p. 112).

Why do science publications use peer review?

Given the imperfections of peer review, one might wonder why it is so widely used. Drummond Rennie, a deputy editor of *JAMA*, suggests that 'peer review represents a crucial democratization of the editorial process, incorporating and educating large numbers of the scientific community, and lessening the impression that editorial decisions are arbitrary' (Rennie 2002, p. 2759). Considering that Rennie has devoted much of his life to editing journals and encouraging the study of peer review (as one of the founding figures of the international congresses on peer review), this seems a rather luke-warm endorsement. Given the resources invested in peer review by publishers and learned societies, and the time devoted to it by academics, one might imagine that it served a higher purpose than merely 'lessening the impression that editorial decisions are arbitrary'. Yet, as we have seen, the formal evidence that peer review 'works' is largely unconvincing. One reason why we have failed to demonstrate conclusively that peer review is effective may be because the term is applied to a wide range of journal processes and is used by different editors to achieve different ends. Unless we can agree on the function of peer review it will be impossible to show either that it is, or is not, fit for purpose (Jefferson *et al.* 2001). It should also be remembered that finding no evidence of an effect is not the same as finding evidence that there is no effect (in other words, all we can say at the moment is that while we do not know whether peer review works, we cannot conclude that it definitely does not work).

One reason why journals continue to invest in peer review may be that most editors, and many authors, can recount anecdotal evidence that peer review improves the quality of submissions. Studies comparing submitted manuscripts with published papers generally show slight but measurable improvements in readability but this may have as much to do with technical editing (i.e. the final process of preparing the paper for publication and applying the journal's house style) as with peer review in the sense of authors responding to reviewers' and editors' suggestions (Wager and Middleton 2002). Peer reviewers and technical editors undoubtedly do detect errors and omissions and may improve the quality of reporting by correcting ambiguities and grammatical errors and by making articles easier to understand. However, most experienced authors can also recount instances when technical editors have changed their meaning or when errors have arisen during the editorial process.

Another reason why peer review persists is that different types of organization (not only journal editors) value it. For example, commercial companies wishing to publish research on their products value the imprimatur of an independent journal. Findings published in peer-reviewed journals are more likely to be respected than findings published by the company itself, which might be dismissed as advertising. Learned societies may also be concerned to preserve their reputation and ensure the quality of research published under their name in their own journals. For this reason they have an interest in using peer review as a quality control mechanism, despite its imperfections.

Peer review has also become embedded into the systems used for academic appointments and for evaluating academic departments to determine future funding. One of the main criteria for getting a job at a university is the number of peer-reviewed papers a researcher has published. Similarly, the number of publications produced by a department will affect how much funding it receives in future (see Box 1). Since peer-reviewed publications have become the currency of academia, any changes to the publishing system are likely to meet with resistance from academics unless the reward system is also changed.

BOX 1 THE USE AND ABUSE OF IMPACT FACTORS

One measure of the influence of a particular piece of research is the number of times it is cited in other articles. Journals can therefore be ranked according to how often, on average, their papers are cited. This figure is called the journal's impact factor; it is calculated from the number of citations to a particular journal in the previous 2 years (appearing in the journals monitored by the *Journal Citation Reports* produced by Thomson ISI) divided by the total number of articles published in the journal in that period. As their inventor, Eugene Garfield, has pointed out, impact factors were designed to compare the overall influence of different journals, not to measure the importance of individual publications (Garfield 2006).

Analyses of the statistics show that it is heavily skewed for most journals, with a small number of papers contributing disproportionately to the average. This average figure is therefore not a reliable way to measure the impact of an individual article, yet it is widely used in this way.

In many countries, the publication records of individual researchers and departments are measured according to the impact factor of the journal in which they were published. For example, a paper appearing in a journal with an impact factor of 12 receives 12 publication points, while one published in a journal with an impact factor of 2 gets 2 points and a certain number of points must be accumulated for a particular appointment, such as a professorship; alternatively, a department's output may be judged according to the number of papers published in journals with an impact factor >5 (or some other arbitrary figure).

This misuse of journal impact factors as a surrogate for evaluating individual publications has resulted in increased competition to publish in journals with high impact factors and reluctance to publish in local, new or electronic journals that are not included in Thomson's *Journal Citation Reports* and therefore lack an impact factor.

The European Association of Science Editors (EASE) has recently issued a statement condemning this practice. They recommend that 'journal impact factors are used only – and cautiously – for measuring and comparing the influence of entire journals, but not for the assessment of researchers or research programmes either directly or as a surrogate' (EASE 2007, p. 99). However, impact factors are widely used across Europe and the USA for evaluating candidates for academic posts and determining the levels of research funding for university departments (EASE 2007).

New developments in peer review

While the principles of obtaining comments from independent experts have remained unchanged since at least the 17th century, the mechanisms for obtaining these opinions changed dramatically in the 20th century with the introduction of fax and then e-mail. The task of identifying potential reviewers has also been revolutionized by bibliographic data bases (such as PubMed) and electronic search engines (such as Google Scholar) making it simple for journal editors to get information about an individual's area of research, publication record and contact details. This has probably globalized many journals' reviewer data bases, since it is now feasible and affordable to obtain reviews from experts virtually anywhere on the planet.

However, although e-mail may have made it possible for editors to contact a wide pool of potential reviewers, and may have trimmed a few days off the process of sending out manuscripts and returning comments, for most journals, peer-review times have not shortened dramatically and authors still often have to wait several months for a decision. For example, in 2006 the average time from submission to acceptance at *JAMA* was about 3 months (Fontanarosa and DeAngelis 2007).

The advent of electronic publishing and journal web sites has reduced lead times (i.e. the time from acceptance to publication) since articles may be published on the web before appearing in print. However, the rate-limiting step in peer review remains the willingness of researchers to perform reviews, their availability and other commitments on their time. Since reviewing for journals is usually unpaid and rarely recognized as a high priority or even as an accepted part of academic workload, it tends to slip to the bottom of the heap, causing delays. Surveys suggest that it takes, on average, about $2^1/2$ hours for an experienced reviewer to review a submission (Yankauer 1990). Black *et al.* (1998) found that review quality increased with the length of time spent on the review up to 3 hours but not beyond that.

A few publishers, realizing that companies often want to publish their research findings quickly, and are prepared to pay for this facility, have established journals that offer rapid review and publication, for a fee. These journals achieve rapid peer review by paying their reviewers and tend to use a smaller pool of reviewers than other journals. However, these reviewers are considered sufficiently knowledgeable to be regarded as the authors' peers even though they may not be the best-known experts in a particular field. Although these so-called 'pay journals' have higher acceptance rates than the most prestigious, traditional journals (sometimes accepting around 80% of submissions), they do not accept all submissions and they also use peer review to improve submissions and detect errors. These differences in the choice of reviewers and in acceptance rates illustrate the difficulty of comparing the quality of peer review between different journals.

The development of the internet has created possibilities for alternative models of peer review. For example, instead of contacting a small number of reviewers, selected for their expertise in a particular topic, web publishing makes it possible to make submissions available to a much wider audience of potential reviewers. The *Medical Journal of Australia* (*MJA*) experimented with posting submitted articles on its web site for a certain period so that anybody could offer comments and suggestions for improvement. However, this different review system was not subsequently adopted by the *MJA* because not enough comments were received and the editors did not feel that this so-called system of 'open review' yielded reviews of equivalent quality and depth compared with the traditional system (Bingham 2003).

The science journal *Nature* also ran a trial of open peer review in 2006 during which authors were asked if they would agree to have their papers posted on the *Nature* web site for open comment at the same time as the papers were being sent for traditional peer review (i.e. before acceptance or rejection) (Anon 2006). Only 5% of the authors of the 1369 papers agreed and, of the 71 displayed papers, 33 received no comments. *Nature* therefore concluded that this system of review was not feasible and reverted to their traditional system. It appears, therefore, that although some journals have experimented with new systems, none has, so far, radically altered their peer-review methods.

Another possibility is to use the conventional system of pre-publication review by a small number of invited reviewers, but then to encourage post-publication comment on published articles. Most journals have traditionally published selected letters from readers, but electronic publishing has increased the potential speed of such responses and the number that can be published. A few journals now offer the possibility not only for readers to comment on a publication but also to annotate it. An example of a new biomedical journal set up with this facility is the Public Library of Science's *PLoS One*.[4]

A great deal of physics research is posted on the preprint server arXiv, which was started in 1991 before any physics journals were available online, and now contains almost half a million electronic articles (see also Chalmers, Chapter 2.2 this volume). Papers on arXiv are not peer reviewed, although they are screened to ensure they are not offensive and are 'at least of refereeable quality'. The server is widely used, with over 20 million full text downloads during 2002, yet, perhaps surprisingly, it has not replaced the traditional physics journals and the great majority of papers that are posted on arXiv are also submitted to peer-reviewed journals. Ginsparg (2003) suggests that the rapid access and basic screening explain why 'expert arXiv readers are eager and willing to navigate the raw deposited material, and greatly value the accelerated availability over the filtering and refinement provided by the journal editorial processes (even as little as a few months later)'. Non-reviewed preprint servers have been proposed for medical research but have not been adopted in the same way as for physics, perhaps because of concerns that poorly conducted medical research or misleading reports could harm patients.

The future of peer review

Alternative methods of publishing, such as physics preprints and the growing number of drug companies that post results of their clinical trials on web sites, may challenge the dominance of the traditional peer-reviewed journal as the primary source of research findings. Non-peer-reviewed methods of publishing research findings may therefore increase in the future.

In the United States a law was passed in September 2007 (coming into effect in 2009) making it compulsory to make results of clinical trials available within a year of the end of a study. This is likely to increase the use of results web sites since it is quicker and cheaper to post results on such web sites than to get them published in peer-reviewed journals. At present (2008), these web sites are not peer-reviewed. If this method of publication becomes more widespread and accepted, it is possible that journals will publish fewer papers describing the results of clinical trials. Publishers may therefore need to rethink their strategies and reconsider the role of their medical journals. Since peer review has not been found to be an effective method of detecting fraudulent data, but may be useful for improving more subjective accounts of the interpretation and implications of new findings, in future it may be reserved for such reviews and syntheses

---

4. http://www.plosone.org.

while primary data may be published in non-peer-reviewed forums. Clinicians and patients will continue to rely on journals to inform them about new research, but the role of journals as the source of primary data may decrease and the emphasis of some journals may change (Wager 2006).

Conclusion

Richard Smith (former editor of the *BMJ*) has written that 'peer review is a flawed process full of easily identified defects with little evidence that it works' (Smith 2006, p. 94). Similarly, Richard Horton (the editor of *The Lancet*) has stated 'We portray peer review to the public as a quasi-sacred process that helps to make science our most objective truth teller. But we know that the system of peer review is biased, unjust, unaccountable, incomplete, easily fixed, often insulting, usually ignorant, occasionally foolish, and frequently wrong' (Horton 2000, p. 148). Yet both the *BMJ* and *The Lancet*, along with tens of thousands of other science journals continue to use peer review to select articles for publication. Smith suggests this is 'because there is no obvious alternative and scientists and editors have a continuing belief in peer review'. He also comments 'How odd that science should be rooted in belief' (Smith 2006, p. 94). Many editors, it seems, accept that peer review is far from perfect yet consider (like Winston Churchill's famous lines about democracy) the alternatives to be even worse.

While electronic communication has revolutionized other aspects of science publishing, it has not fundamentally changed the peer-review process, although it has probably speeded it up a little. A handful of journals have experimented with different methods of peer review but none has found them to be workable. Even if new methods evolve in the future, acceptance and adoption may be hampered by the fact that peer-reviewed publications have become the 'currency' by which academic research output is measured, and, unless this system is changed authors may be reluctant to risk alternative methods of publishing.

Ironically, the explosion of information available on the web may actually increase the desire of many readers for systems (such as peer review) that can filter information and assess its quality and reliability. The demand for evaluation of information on the internet may therefore increase and might, perhaps, spawn new variations on peer review and journals will continue to be regarded as more trusted sources of information than other sites. It therefore seems likely that, despite its obvious shortcomings, science journals will continue to invest in peer review in a form that would be largely recognizable to scientists from 300 years ago.

■ REFERENCES

ALPSP/EASE Survey (2001). *ALPSP/EASE Current Practice in Peer Review: Results of a Survey Conducted in Oct/Nov 2000*. Available at: **http://www.alpsp.org/ForceDownload.asp?id=140**.

Anon (2003). Coping with peer rejection. *Nature*, **425**, 645.

Anon (2006). Overview: *Nature*'s peer review trial. *Nature*. Available at:
http://www.nature.com/nature/peerreview/debate/nature05535.html.

Bingham, C. (2003). Peer review on the internet: are there faster, fairer, more effective methods
of peer review? In: *Peer Review in Health Sciences*, 2nd edn (ed. F. Godlee and T.O. Jefferson),
pp. 277–93. BMJ Books, London.

Black, N., van Rooyen, S., Godlee, F., Smith, R. and Evans, S. (1998). What makes a good
reviewer and a good review for a general medical journal? *Journal of the American Medical
Association*, **280**, 231–3.

Burnham, J.C. (1990). The evolution of editorial peer review. *Journal of the American Medical
Association*, **263**, 1323–9.

Dickersin, K., Fredman, L., Flegal, K.M., Scott, J.D. and Crawley, B. (1998). Is there a sex bias in
choosing editors? *Journal of the American Medical Association*, **280**, 260–4.

EASE (European Association of Science Editors) (2007). EASE statement on inappropriate use
of impact factors. *European Science Editing*, **33**, 99–100.

Fisher, M., Friedman, S.B. and Strauss, B. (1994). The effects of blinding on acceptance of
research papers by peer review. *Journal of the American Medical Association*, **272**, 143–6.

Fletcher, R. and Fletcher, S. (2003). Effectiveness of peer review. In: *Peer Review in Health Sciences*,
2nd edn (ed. F. Godlee and T.O. Jefferson), pp. 68–9. BMJ Books, London.

Fontanarosa, P.B. and DeAngelis, C.D. (2007). Thank you to *JAMA* peer reviewers and authors.
Journal of the American Medical Association, **297**, 875.

Garfield, E. (2006). The history and meaning of the journal impact factor. *Journal of the American
Medical Association*, **295**, 90–3.

Ginsparg, P. (2003). Alternatives to peer review II: can peer review be better focused? In:
Peer Review in Health Sciences, 2nd edn (ed. F. Godlee and T.O. Jefferson), pp. 312–21.
BMJ Books, London.

Godlee, F. and Dickersin, K. (2003). Bias, subjectivity, chance, and conflict of interest in editorial
decisions. In: *Peer Review in Health Sciences*, 2nd edn (ed. F. Godlee and T.O. Jefferson),
pp. 91–117. BMJ Books, London.

Godlee, F., Gale, C.R. and Martyn, C.N. (1998). Effect on the quality of peer review of blinding
reviewers and asking them to sign their reports: a randomized controlled trial. *Journal of the
American Medical Association*, **280**, 237–40.

Horton, R. (2000). Genetically modified food: consternation, confusion, and crack-up. *Medical
Journal of Australia*, **172**, 148–9.

Jefferson, T., Wager, E. and Davidoff, F. (2001). Measuring the quality of editorial peer review.
Journal of the American Medical Association, **287**, 2786–90.

Jefferson, T.O., Alderson, P., Wager, E. and Davidoff, F. (2002). Effects of editorial peer review:
a systematic review. *Journal of the American Medical Association*, **287**, 2784–6.

Kronick, D.A. (1990). Peer review in 18th-century scientific journalism. *Journal of the American
Medical Association*, **263**, 1321–2.

Lloyd, M.E. (1990). Gender factors in reviewer recommendations for manuscript publication.
Journal of Applied Behavior Analysis, **23**, 539–43.

Lock, S. (1991). *A Difficult Balance: Editorial Peer Review in Medicine*. BMJ Books, London.

MEPS (*Marine Ecology Progress Series*) (2000). Theme issue on peer review. *Marine Ecology Progress
Series*, 192 (31 January). Available at: http://www.int-res.com/articles/theme/riisgard.pdf.

Nylenna, M. and Horton, R. (2006). Research misconduct: learning the lessons. *The Lancet*,
368, 1856.

Peters, D.P. and Ceci, S.J. (1982). Peer-review practices of psychological journals: the fate of published articles, submitted again. *Behavioral and Brain Sciences*, **5**, 187–95.

Petit-Zeman, S. (2003). Trial by peers comes up short. *The Guardian*, 16 January, p. 10.

Rennie, D. (2002). Fourth International Congress on Peer Review in Biomedical Publication. *Journal of the American Medical Association*, **287**, 2759–60.

Smith, R. (1994). Conflict of interest and the *BMJ*. *British Medical Journal*, **308**, 4–5.

Smith, R. (2006). *The Trouble with Medical Journals*. Royal Society of Medicine, London.

Wager, E. (2006). Publishing clinical trial results: the future beckons. *PLoS Clin Trials*, **1**, e31 (doi 10.1371/journal.pctr.0010031).

Wager, E. and Middleton, P. (2002). Effects of technical editing in biomedical journals: a systematic review. *Journal of the American Medical Association*, **287**, 2821–4.

Wager, E., Parkin, E. and Tamber, P.S. (2006). Are reviewers suggested by authors as good as those chosen by editors? Results of a rater-blinded, retrospective study. *Biomed Central*, **4**, 13.

Wennerås, C. and Wold, A. (1997). Nepotism and sexism in peer-review. *Nature,* **387**, 341–3.

Wood, F. and Wessely, S. (2003). Peer review of grant applications: a systematic review. In: *Peer Review in Health Sciences*, 2nd edn (ed. F. Godlee and T.O. Jefferson), pp. 14–44. BMJ Books, London.

Yankauer, A. (1990). Who are the peer reviewers and how much do they review? *Journal of the American Medical Association*, **263**, 1338–40.

■ FURTHER READING

- Godlee, F. and Jefferson T.O. (eds) (2003). *Peer Review in Health Sciences*, 2nd edn. BMJ Books, London. An edited collection covering all aspects of peer review in health sciences and summarizing the evidence available from research.

- Hudson-Jones A. and McLellan, F. (eds) (2000). *Ethical Issues in Biomedical Publication*. Johns Hopkins University Press, Baltimore, MD. An edited collection considering the ethics of peer review (including ethical issues surrounding electronic publishing) with thoughtful contributions from several leading journal editors.

- Smith, R. (2006). *The Trouble with Medical Journals*. Royal Society of Medicine, London. A personal view from the former editor of the *BMJ* focusing on the influence of the pharmaceutical industry on medical journals but also addressing some of the shortcomings of peer review; aimed at the general public and very readable.

- Wager, E., Godlee, F. and Jefferson, T. (2002). *How to Survive Peer Review*. BMJ Books, London. A practical handbook for authors and reviewers, distilling some of the information from Godlee and Jefferson (2003).

■ USEFUL WEB SITES

- **Committee on Publication Ethics (COPE): http://www.publicationethics.org.uk/.** The COPE web site includes short accounts of cases discussed by the committee and therefore gives insights into the problems facing journal editors. The web site also contains flowcharts to guide editors when they encounter possible cases of misconduct.

- **International Committee of Medical Journal Editors (ICMJE):** http://www.icmje.org/. The ICMJE 'Uniform Requirements' remain the most widely respected guidelines about medical publishing for authors and editors. They consider issues such as authorship, the role of reviewers and redundant publication.

- *Nature* **peer-review debate:** http://www.nature.com/nature/peerreview/debate/index.html. An interesting (although ultimately unsuccessful) example of a journal experimenting with an alternative form of peer review. The *Nature* peer-review trial is described on this site and there are also general articles about peer review from a number of experts.

- **Open access overview:** http://www.earlham.edu/~peters/fos/overview.htm. An account of the 'open-access' debate.

4.2

Controversy and consensus

Jeff Thomas

Controversy has always been a driving force for the advancement of scientific knowledge. Disputes have often been contained within the narrow boundaries of academia, until hard-won consensus occasionally achieves a settlement of sorts. The late John Ziman (2000) pointed out that the simultaneous promotion of both controversy and agreement is a distinctive feature of scientific progress, reflecting (respectively) the principles of 'organized scepticism' and of 'universalism' that (as Robert Doubleday argues in Chapter 1.2 in this volume) still have a slender bearing on how modern science is practised. But to the contemporary view, at least through the lens of news media discourse, science is at least as rich in discordance and uncertainty as it is in the settled harmony of consensus.

This chapter argues that the changing nature and prominence of contemporary controversies in the public sphere provides both threats and opportunities for working scientists. The range of behavioural practices that comprise the communication 'tool-bag' of contemporary scientists is changing, requiring of scientists attributes and sensitivities that have been traditionally neglected in scientific training and practice. The aim here therefore is not to ambitiously develop a generalized typology of controversy as others have done (see Martin and Richards 1994), or indeed to explore how controversies arise, mature and fester, or achieve (or fail to achieve) closure (see Nelkin 1994), but to say something about the novel ways in which contemporary scientists do and might behave amidst scientific controversy, and identify new communication practices that are likely to be increasingly encouraged and rewarded. A key issue is whether such a likely change in these practices would be beneficial. The chapter begins by looking at a couple of controversies prevalent at the time of writing (2007–2008), which help to frame what follows.

From the particular to the general

In November 2007 the media's attention was temporarily focused on a scare in reverse —the possibility that being mildly obese might bring associated health benefits. This analysis, published by a premier research institution (the Centers for Disease Control and

Prevention in Atlanta, GA, USA) and prominent as a front-page article in the UK's *Independent* newspaper (Usbourne 2007), was based on the analysis of 'decades of data by federal researchers'; it reported that, on a population basis, modestly overweight people have a lower death rate than those who are underweight, obese or of normal weight (see Flegal *et al.* 2007). This research runs counter to the bulk of preceding published findings which, in the UK at least, swiftly became the foundation for public health policy extolling the virtues of tight control of body weight, with an unqualified emphasis on the health benefits of dietary restraint. *The Independent* article also reported that being overweight may be associated with improved chances of recovery from surgery and infections and improved prognosis for disease.

Key elements evident in this debate have a wider applicability. First, no opportunity was lost for the reporting of instantaneous condemnations by contrary experts. Thus the same *Independent* article continues: 'It's just rubbish' fumed Walter Willet, professor of Epidemiology and Nutrition at the Harvard School of Public Health, adding that 'it is just ludicrous to say that there is no increased risk of mortality from being overweight' (quoted in Usbourne 2007). Disputes between experts have become very much a feature of public controversy, fuelling a greater awareness on the public's part of the limitations of experts' knowledge and of the likelihood that (in the words of Martin and Richards 1994, p. 507) 'purportedly "disinterested" advice may be influenced by professional, economic, or political considerations'. (Indeed, a subsequent letter to the editor of *The Independent* drew attention to the need for research conducted by 'respected scientists who are not financed in any way by food manufacturers' (Abbott, 2007).) It is not coincidental that this apparent erosion of belief in the 'neutral, disinterested and objective expert'—a self-serving view long promulgated by scientists themselves—is associated with an increased eagerness for greater public involvement in deciding policy about science related issues—a movement reflected on, for example, by Stilgoe and Wilsdon (2009).

Secondly, this controversy, like many others, reveals an underlying science-in-the-making (Latour 1987) that is not only incomplete but complex, and which makes both effective communication and comprehension problematic. For example, Flegal *et al.*'s (2007) data relate to *deaths* associated with weight differences; they convey little about susceptibility to disease or disability. And the benefits that the researchers claimed were evident only in those *mildly* overweight—with a body mass index (BMI) of between 25 and 30. Being genuinely obese *raises* risks, giving lie to the *Independent*'s prefacing headline that trumpets this research, 'Now doctors say it's good to be fat'.

It is still the case that the risks of developing diabetes, for example, increase proportionately with body weight. Flegal's team, who had been researching these data over many years, drew attention to factors such as cholesterol levels, blood pressure and other obesity-related conditions that influence risk for any one individual. Moreover, the analysis reviewed survival data only on 'healthy' individuals who were non-smokers—thin individuals as a group were excluded on the grounds that they may have been unwell in some unknown respect. In some news channels (e.g. in radio interviews) at earlier stages of her research analysis, Flegal has proved more able to emphasize the limits and uncertainties of her data, pointing out 'we don't like to really conclude there's a protective effect (i.e. from being mildly overweight) because there are uncertainties in these methods' (ABC National Radio 2006). Such important notes of caution were seemingly lost in some of the accompanying newsprint coverage: for example, an editorial piece entitled 'Gobble, Gobble' (*Washington Times* 2005), used earlier findings from Flegal and her team to gave licence to its readers, on the eve of the North American festival of Thanksgiving, to 'pass the gravy and the potatoes and an extra helping of turkey' with a guilt-free conscience.

Of course, this research touches on a health issue of enormous sensitivity and concern in the USA—and increasingly the UK—which explains why the research data have enjoyed such a high public profile, especially in the blogosphere and on partisan web sites that are increasingly evident in public health controversies. The number and range of contributions to this web 2.0 discourse is impressive—an uncontrolled explosive fragmentation typical of topics of this significance and newsworthiness, with the story moving well beyond the influence and comfort zone of the originating scientists.

The role of scientists in public controversies

Controversy is a pre-eminent feature of contemporary science for a further reason; controversial scientists abound, offering a mix of self-promoting public utterances, behaviours and misdemeanours that ensure science is seldom far from the public spotlight. Of course, professional vanity and self-promotion is a long-established and valued tradition in the scientific community, for some at least. However, the visibility of this very human face of science now seems especially significant, if only because of the multiplicity of channels through which fame and notoriety can now be claimed.

Again, a recent example tellingly makes a broader point. At the time of writing (2007–2008), controversy surrounds the celebratory events associated with the launch of James Watson's (a co-discoverer of the structure of DNA) most recent popular science book *Avoid Boring People* (Watson 2007) and in particular a widely publicised UK promotional tour. Here he let slip in an otherwise run-of-the-mill interview (see Hunt-Grubbe 2007) the opinion that 'all our social policies are based on the fact that their [i.e. black Africans'] intelligence is the same as ours—whereas all the testing says not really'. As fallout from the ensuing debacle, the UK promotional tour was brought to an unhappy end, as was his chancellorship of Cold Spring Harbor Laboratory (McKie 2007).

As Kohn (2007) points out, these public pronouncements, with pronouncement swiftly followed by contradictory public apology, contrasts with the more measured tones of his formal writing of the book, which ends with a more circumspect pronouncement that 'there is no reason to anticipate that the intellectual capabilities of people geographically separated in their evolution should prove to have evolved identically'. Perhaps here Watson is disregarding of his own homily in *Avoid Boring People* to 'always remember your intended readers' (Watson 2007, p. 238), forgetting that the interview was far from an intimate *tête-à-tête*. An explosion of rage and offence followed Watson's pronouncement, with critics inside and outside science competing for adjectives of condemnation. Thus a UK government minister (David Lammy, the Skills Minister, quoted in McKie 2007) bemoaned that 'it is shame that a man with a record of scientific distinction should see his own work overshadowed by his own irrational prejudices'. Some commentators urged that the views of Watson would be all the more swiftly consigned to the rubbish bin if, in the spirit of free speech, his argument continued to be freely broadcast and hence more openly confronted (Porter 2007).

But the dominant voices, from established scientists and social commentators, via blogs and more established channels alike, were hostile and condemning. Watson's long-time opponent Steven Rose, for example, argued for a necessary curtailment of free speech in this instance. Just as with 'hate speak' and shouting 'fire' unwarrantedly in a crowded theatre he saw this necessary banishment as hastening Watson's despatch to the footnotes of the extended controversy on scientific racism (Rose 2007). By this logic, the cancellation of a planned Science Museum event was justified on the grounds that his remarks went 'beyond the point of acceptable debate'. Several prominent UK scientists bemoaned this sequence of events; Richard Dawkins (quoted in McKie 2007) asserted that:

... what is ethically wrong is the hounding, by what can only be described as an illiberal and intolerant 'thought police' of one of the most distinguished scientists of our time, out of the Science Museum and maybe out of the laboratory he has devoted much of his life too, building up a world-class reputation.

What was often lost (though not from Rose's arguments) was the thinness of Watson's science—this example of pseudoscientific racism is most striking because of the absence of tenable evidence that could be mustered in support.

Though pathological in scale, this ill-tempered controversy hints at something dark and unappealing about the conduct of science and the social sensitivities of scientists. With operatic intensity, egos come writ large and hubris and conflict abound. The notion of a 'wise scientific community' and of consensus seems in short measure; indeed Watson had earlier written:

One could not be a successful scientist without realising that, in contrast to the popular conception supported by newspapers and the mothers of scientists, a goodly number of scientists are not only narrow-minded and dull, but also just stupid.

Watson (1968, p. 18)

As seems often the case, it is through the medium of journalistic coverage (and popular science books; see Thomas 1999), that the inner and seemingly uncomfortable workings of science are revealed in exaggerated emphasis.

Fortunately, egos and hubris on this scale are exceptional, but their effect on public perception is disproportionate and unfortunate, as are the messy episodes of science-in-the-making characterized earlier. What follows in the remainder of this chapter suggests that it need not always be so; that, for better or worse, the stage has never been more favourably set for playing out controversies in which scientists themselves have a more active and participative role, through their greater awareness of modern communication practices. The focus here is on pragmatic and evolutionary change; others argue for intellectual and structural developments much greater in scope, as with the construction of socially robust knowledge advocated by Michael Gibbons and colleagues (Nowotny *et al.* 2001). In part, the emerging paradigm described here is linked with the increased emphasis on public consultation and dialogue—and all the challenges inherent in putting such grand intentions into action that Irwin (2009) has highlighted.

Scientists, persuasion and public dialogue

In the UK, the bovine spongiform encephalopathy (BSE) episode, proved an especially influential controversy, in that it demonstrated science communication in disarray (see Jasanoff 1997; Irwin 2009). More recently, the UK's *GM Nation?* public consultation exercise has achieved a comparable iconic status, but again more out of notoriety. For many, it proved a consultation exercise that we can ill-afford to replicate if engagement with publics is to be meaningful and productive. At a time when the government's rhetoric presents an image of a more polished and professional approach to consultation and dialogue in the UK (Office of Science and Innovation 2006), it is instructive to ask what went wrong, not with negative intent but more to illustrate opportunities for good practice missed.

The commercialization of genetically modified (GM) crops has proved a major area of controversy, with the UK government even now persisting with the view that (in the words of the recently retired government's chief scientific adviser) GM crops are 'essential to deal with an ever expanding global population and diminishing water supplies' (Professor Sir David King, quoted in Anon 2007c). But opposition to GM technology from about mid-1998 onwards in the UK seemed widespread and deep-seated, so the decision to hold a national dialogue on GM (the *GM Nation?* debate) appeared both laudable and risky—in the sense that a 'favourable' outcome (from the government's perspective, who were broadly supportive of GM technologies) was far from assured.

The *GM Nation?* consultation exercise doubtless provided a valuable opportunity for scientists to engage in public dialogue. The scale and type of meetings conducted nationwide in the UK was unprecedented—more than 600 in number and ranging from half-a-dozen highly orchestrated regional meetings, with an overall attendance of about 1000, to informal, small-scale meetings attracting as few as 30. The majority of them required an input from what was from the beginning unfortunately and simplistically labelled 'the two sides of the debate'. Those generally classifiable as the opponents of GM technology (e.g. non-governmental organizations (NGOs) such as Friends of the Earth) were seemingly more prominent and adept in their contributions to public debate than those—notably the scientists—whose role was to argue in favour of the adoption of GM technology. Indeed, a number of public meetings went ahead with no scientific spokesperson defending GM technology (see Horlock-Jones *et al.* 2004). When these two sides of the debate were present, the anti-GM NGOs were not only more assiduous and better organized, but their concerns about GM were better articulated and received than counter arguments of the pro-GM scientists.

Irwin (2009) mentions how the structure of the *GM Nation?* debate tended to isolate technical aspects from broader social issues; very often, the contribution from scientists failed to move beyond the narrow confines of the comfort zone of narrowly interpreted technical content. This hesitancy on behalf of scientists is likely to be both strongly felt and widespread. A MORI (2001) poll conducted in 2000 on the role of public scientists in public debate reported that while 20% of scientists polled felt themselves 'very well equipped' to communicate the scientific *facts* of their research—itself a modest enough figure—only 10% of those polled felt themselves very well equipped to communicate the social and ethical implications of their work. Oddly, most scientists polled in the same survey feel their profession should have the main responsibility of communicating the ethical and social implications of their research to non-specialist publics.

If scientists approached the *GM Nation?* exercise as an opportunity for persuasion—that in true 'deficit model' thinking, 'knowing the facts' would persuade participants to a more pro-GM stance—the outcome strongly suggested otherwise. On the evidence provided by a small control group (see below), the more individuals became engaged with GM issues, actively taking steps to find out more about them, the more resolved their attitude generally became, whether pro or anti GM. While some of the potential benefits from GM were recognized, notably the potential medical benefits, people expressed increasing concern about the risks associated with GM. The area that proved most influential in this hardening of opinion was the imagined long-term effect of GM on human health. Where research and discussion revealed areas of continuing uncertainty, touching on the many questions about GM that remain unanswered, or are unanswerable, expressions of unease about moving forward to commercial exploitation were reinforced.

What was clear by way of outcome from the *GM Nation?* debate was that the great majority of those who became involved in the debate were in one way or the other 'uneasy' about GM, with feelings that most often sat somewhere within the spectrum of 'suspicion and scepticism, to hostility and rejection' (*GM Nation?* 2003). If ever consultation was naïvely viewed as a means of 'settling' controversy, such hopes were clearly unfulfilled, contributing no doubt to the perceived 'messiness' of the exercise.

An adequate data set?

Just how discerning was such monitoring of public opinion? The self-selecting status of those 36,000 or so who participated in the 'open debate'—many of these respondents had attended one or other of the meetings previously described—was a point of concern to those compiling the data (*GM Nation?* 2003). About half of such 'open' responses received came in by conventional mail and half by filling in the questionnaire on the *GM Nation?* web site. More than likely, the invitation to send in opinions about GM would be grasped by those with existing, often strong opinions about GM, as opposed to what could be termed in such circumstances the less opinionated 'silent majority'. One of the checks applied by the organizers of the *GM Nation?* debate was to run the closed 'control' group just mentioned, of just 78 randomly selected people, chosen to reflect as far as was possible the make-up of the general population, and who were asked the same 13 questions responded to by the much larger sample group that took part in the open debate. Opinion within this control group was sampled again after a 2-week period of discussion and self-directed research, to see if opinion had shifted over that period.

The authors of the *GM Nation?* report claimed that responses to the same 13 questions from the open and closed groups were broadly in agreement, some minor differences apart. This was crucial evidence to support their conclusion that members of the public, sampled in the open debate, were not 'a completely different audience with different values and attitudes from an unrepresentative activist minority'. And yet Campbell and Townsend (2003) have analysed the two sets of data in more detail and noted what they term startling differences between the two groups for more than half the questions, which were not made evident in the *GM Nation?* report. For example, in response to the question 'I would be happy to eat GM foods' just 8% of those in the open-debate group agreed; in the smaller control group, 35% of the respondents answered positively. Campbell and Townsend (2003) come to the reasonable conclusion that the *GM Nation?* authors' analysis of the representativeness of views expressed in the open debate should be more firmly challenged. Further, these authors are critical of 'vague and leading questions' used in the debate, as well as the lack of contextualization of the questions; they feel (as do a number of other authors) that the degree of non-acceptance of the GM technology at issue was over-played.

Quite why the 'scientific voice' in this controversial area seemed oddly muted and discordant is uncertain, but the lack of consensus and commitment amongst scientists was a contributing factor. The key point here is not to deny the genuine misgivings about GM foods in the population at large that the *GM Nation?* exercise uncovered. Rather, it is to ask what type of scientific representation might be increasingly evident in future debates—and also to urge that greater thought and effort be applied to finding out more about attitude formation and the moulding of 'public opinion' (see Thomas 1997). As the next example illustrates, opportunities for scientists to engage themselves more extensively and intrusively in public debate abound, though effective involvement seems all the more likely to require them to work with—rather than against—the grain of long-established news media practices.

Stem cells, persuasion and 'media savvy' scientists

The House of Lords (2000) report saw public consultation as in some sense *validating* good policy-making. In UK, policy-makers are still drawn to the ideal (if not always the practice) that a level of 'social acceptance' is required for scientific and technical innovation. In the field of reproductive biology for example, the Human Fertilisation and Embryology Authority (HFEA), following the legislative framework established by the 1990 Act of the same name, have a licensing role in approving new forms of research in pioneering areas such as the use of stem cells and somatic cell nuclear transplantation, more commonly termed 'cloning'. The HFEA put great emphasis on the importance of a significant measure of public support for the guidelines they create. Public non-acceptance is one reason why the creation of genuine hybrid embryo—of the type that might be formed, for example, by the fertilization of an animal egg with human sperm—continues to be a step too far, at least at the time of writing (2007–2008). In reality, the biological value of any such a hybrid is so questionable—and the associated ethical doubts so serious—that at present the procedure lacks credible scientific champions, just as do chimeras, which are embryos that would be formed by mixing animal and human cells.

In 2007, two groups of UK scientists—one led by Dr Stephen Minger of King's College, University of London and the other by Dr Lyle Armstrong at the North East England Stem Cell Institute in Newcastle—applied to the HFEA for permission to create cytoplasmic hybrids, now colloquially (and unattractively) termed cybrids. In doing so, these researchers were seeking to get round the problem of the large numbers of eggs that are required to derive cloned embryonic stem cells. Many scientists feel that these types of cells hold the best hope for the treatment of progressively debilitating diseases—for example motor neuron disease. Up until that time in the UK 'spare' eggs were usually donated for such purposes by women undergoing *in vitro* fertilization (IVF) treatment. Given the inconvenience (and to some measure, the health risks) involved in this procedure, the volunteered supply of such eggs is hugely outweighed by the demand for them from scientists aiming to create new stem cell lines for therapeutic purposes. One way round this bottleneck is to use animal eggs—from cattle or rabbits for example—rather than humans for the creation of such stem cell lines. First, the nucleus of these animal cells is removed prior to the introduction of a donor human nucleus; what is thereby created is a mix of human DNA (contained within the introduced human nucleus) and a very small amount (comprising about 0.1%) of animal DNA located in the cytoplasm of the recipient egg. (It is the pre-dominance of the one (human) type of DNA that distinguished such an embryo from a true hybrid.) Such embryos are then artificially grown in the lab for no more than 5 or 6 days and then destroyed as the stem cells they contain are extracted and cultivated.

Should such procedures be permitted? The initial skirmishes suggested not. In December 2006, draft legislation had been produced in the form of a White Paper by the Department of Health on behalf of the UK government, outlining intended guidelines for future HFEA policy, which proposed that the use of cytoplasmic cell hybrids be prohibited—as part of a 'blanket ban' on all reproductive techniques involving hybrids and chimeras. Part of the justification for this initial prohibition on the creation of cybrids was the HFEA's preliminary assessment of the public non-acceptability of such work. A spokesperson at the time confidently asserted that 'previous reviews in this area show an on-going and widespread support for a ban on creating human–animal hybrids and chimeras for research purposes' with the Health Minister of the time (Caroline Flint) talking of being sensitive to 'the moral compass of how these technologies are used' (both quotes from Roberts 2007)

An argument from a science perspective

In immediate response from those in the UK keen to see cybrids given the go-ahead, a picture was presented of a strong consensus on the point within the UK scientific community. In reality, scientists in the UK, as elsewhere, rarely speak with one voice on stem cell use in general and the importance of embryos as a source in particular. There is a continuing and sometimes ill-tempered spat about the optimal source of stem cells, though in the UK it has little of the intensity and influence evident in the USA (e.g. see Herold 2006). But in the UK in the early months of 2007, a united front of scientific opinion was mobilized with the aim of reversing the proposed ban on cybrids—and to attempt to do so via news media.

This was a widespread and well-orchestrated promotional campaign. For example, an open letter directed to the then Prime Minister (Tony Blair) urging that cybrid research be allowed was signed by 223 medical research charities and patient organizations (reported in Connor 2007). A further letter published in one of the UK's elite newspapers (*The Times*) was signed by 45 leading scientific figures, including the heads of the most august professional bodies (notably the Royal Society) and the funding bodies responsible for research area (the Medical Research Council) (see Henderson 2007b). Scientists argued that not to proceed with the proposed research would be 'an affront to patients' desperate for therapy (quoted in Henderson 2007a). Patients' groups, representing those affected by conditions such as Alzheimer's, Parkinson's and motor neuron disease, were especially vocal. For example, Belinda Cupid of the Motor Neuron Disease Association said 'the case for the use of human–animal hybrid embryos in stem cell research is compelling as it holds the potential to save lives (quoted in Macrae 2007). On this type of evidence, those who opposed the cybrids ban were able to claim 'people are overwhelmingly supportive of allowing carefully regulated hybrid and chimera embryo research' (Science Media Centre 2007).

Essential to the framing of the pro-science case was the involvement of the Science Media Centre (SMC), a NGO that seeks to 'promote the voices, stories and views of the scientific community to the national news media when science is in the headlines'.[1] This they did adeptly in the context of the cybrids controversy, for example by producing 'briefing notes' for journalists, mainly comprising readily quotable remarks from scientists with concerns about the proposals of the draft bill. The aim of the SMC is to increase the effectiveness of scientists' communication; its director (Fiona Fox) thinks that 'science is about 10 years behind other areas of press relations in terms of playing the media game'. And yet the SMC clearly feel that science journalists are key allies; 'In almost 5 years of putting scientists in front of the national news media on some of the most controversial issues of our time, we can still count the number of bad experiences on one hand' (both quotes from Bithell 2007).

Indeed, the SMC's orchestration could not have happened without the support of campaigning scientists. The doyen of stem cell researchers, Professor Ian Wilmut, in charge of the Roslin Institute team that created Dolly the sheep, weighed in with the view that 'The opportunity to use animals eggs in research will provide an important chance to understand and treat inherited diseases such as motor neurone disease' (quoted in Fleming 2007). And yet surprisingly Wilmut's enthusiasm for such an approach was soon to diminish; by November of the same year he announced that he was to switch to newly emerging techniques, then (as now) gaining in success and support, to create stem cells from skin cells (Topping 2007). This abandonment was described in this news report as 'a blow to scientists who believe that the use of embryos to create stem cells is the best way to develop treatments for serious medical conditions such as stroke, heart disease and Parkinson's disease' (Topping 2007). It was fortuitous that this pronouncement came at this time and not in the midst of the preceding campaign to reverse the proposed ban.

The scientists with most at stake were in the forefront of the campaign. Stephen Minger talked in what might seem an overly optimistic tone of 'our hope that these stem cells will be used to develop new therapies and which could give hope to patients where we currently have little to offer' (Science Media Centre 2007). Dr Minger was a key figure in a SMC-sponsored press conference, described by those in attendance as a 'combative' forum (Coghlan 2007); central to his task at the press conference was that of persuasion and the provision of information. His stated belief (Minger 2007) was that once non-scientists were made aware of the scientific case for the research, in particular the possible medical benefits, they would move from anxiety and hostility to qualified acceptance.

1. See their web site at http://www.sciencemediacentre.org/.

It follows that opposition to the proposed development was generally seen as 'being based on misinformation' (Shaw, quoted in Roberts 2007). Indeed, overcoming expressions of disgust—usually rather dismissively termed 'the yuk factor'—was therefore essential to the campaign. Many scientists expressed their concern that the initial White Paper had been 'more swayed by the yuk factor and by those opposed to scientific progress than by common sense and real understanding of the issues' (Lovell-Badge, quoted in Science Media Centre 2007).

The tendency of these scientists was to dismiss concerns about the 'unnaturalness' of the proposed procedure as irrational and 'unscientific'; proponents of the ban took the view that it reflected a valid, if inarticulate, concern about the direction of modern science and about the manipulation of nature. But the debate that the affected scientists helped frame, as conducted both through conventional news media and the blogosphere, rarely focused on the correctness or otherwise of the 'naturalistic fallacy'—rather it was the promise of medical cures for feared diseases. Supporters of the ban, including those marginalized few 'scientific voices' (for example, see Human Genetics Alert (HGA) 2007) pointed to what they saw as the different worldview of scientists and citizens with respect of the species barrier, in the following terms. Scientists, especially those of a reductionist persuasion, stress the similarities and continuity between different living types, most evident at the molecular level, in a common genetic language. From the non-science perspective, species are very different—'distinct, integrated wholes, different "kinds"'. On this basis, 'barriers' between species have to be respected. Thus, part of the case offered by those supportive of the ban was scientific in nature and put in terms that few scientists would dispute. While graciously conceding that the scientific materialistic world view 'is correct, at one level', HGA continued:

> However, where it fails is in providing any satisfactory account of how assemblages of very similar molecules come to produce species as obviously different as rabbits, cows and humans. There is as yet no good theory of the generation of 'emergent properties' in complex biological systems
>
> Human Genetics Alert (2007)

Uncertainty about the effects of genetic 'mixing' was highlighted, but in this context to little effect, and indeed given the genetic make-up of cybrids (i.e. >99% human), this was perhaps a less than ideal example to make the point.

Such uncomfortably challenging lines of argument, from a self-styled science-based organization not opposed in principle to research on embryos or embryonic stem cells, addressed an agenda quite different from that set by lobbying stem cell scientists. Perhaps surprisingly, given the desire to maintain balance in reporting (see Allan 2009) such arguments failed to find a place within news media coverage, as the therapeutic 'frame' of 'using science to fight disease' often took precedence. The HGA analysis reflects how narrowly framed the public discussion proved; in their words:

> . . . that the public is presumably supposed to judge whether these experiments are worth-while simply on the basis of trust in famous scientists . . . it is sadly too typical of the debate in this country on stem cell research, they will have signed the letter on the basis of their general support for scientific research and their disdain for what they see as woolly-headed and probably religiously-based ethical concerns
>
> Human Genetics Alert (2007)

The HGA also sought to enumerate a number of serious scientific defects of the proposed experiments, questioning their scientific usefulness, but though well-argued, the complexity and high level of scientific analysis inherent in their arguments resulted in little impact.

Part of the adroitness of the scientific case was to frame supporters of the cybrids ban as strongly allied to 'pro-life' and religious groups. Those arguing for the perpetuation of the ban on cybrids were indeed an *ad hoc* coalition of groups that found it difficult to speak with a united or authoritative voice, with arguments that sometimes lurched dramatically from science towards often over-stated ethical concerns. For example, Josephine Quintaville, of the campaign group Comment on Reproductive Ethics offered the perspective that: 'reproduction with animals has been taboo since the beginning of recorded time and that taboo has remained with us for a reason' (quoted in Macrae 2007). In the same article, Anthony Ozimic of the Society of Unborn Children said there were: 'grave ethical and moral objections' to the research saying that 'all the evidence suggests that these embryos are essentially human' and 'yet they will be cannibalised and killed for their stem cells'; he added that patients with degenerative diseases were being exploited and 'they and their families are being sold lies and false hopes by the profit-hungry biotech companies'. The fact that many of the reported objections came from 'pro-life' and/or religious groups, most often opposed to *any* form of embryo research, made it all the easier for such dissent to be marginalized and characterized as unrepresentative of today's apparently secular UK society.

Another neat example of effective lobbying was effective political framing, for example by ensuring that the issue was seen as a broader 'test of the Government's commitment to science'. The campaign benefited from being headed by a politician (and qualified doctor) closer to government circles than the scientists themselves. This was Liberal Democrat MP Dr Evan Harris, who as a member of a political party in opposition to the government was sympathetic to the cause of gathering arguments to undermine government pronouncements. He drew attention to:

. . . an unprecedented show of unity and support from scientists, clinicians, ethicists and patient groups [showing] how concerned people are about this vague, ill thought-through proposal from the Government

quoted in Anon (2007a)

Harris' political leverage was important in establishing a House of Commons Select Committee to assess the issue—a parliamentary grouping charged with scrutinizing intended government policy. It was perhaps predictable that the pronouncement of the Select Committee would be condemning, and so it proved: 'we have found the Government's published proposals for future regulation in this area to be unacceptable and potentially harmful to UK science' (House of Commons 2007, p. 5), emphasizing that the proposed ban 'would undermine the UK's leading position in stem cell research and the international reputation of science in the UK' (House of Commons 2007, p. 54). This episode was portrayed as the litmus paper test of the government's willingness to prioritize research into incurable diseases—their aim was to see science 'win through' on this key battleground and ensure (as the headline in *The Times* of the day expressed it) that the 'white coats defeat the grey suits' (Henderson 2007c).

The SMC-backed campaign also expertly sought to maximize any divisions of opinion within the UK government. Soon after the initial release of the White Paper, the then Chief Scientific Adviser to the government, Professor Sir David King, publicly backed the use of cybrids, as did the Prime Minister Tony Blair and the section of the government at that time most concerned to see the commercial gain from stem cell work realized—the now defunct Department of Trade and Industry. But the Chief Medical Officer (Professor Sir Liam Donaldson) and a former Health Minister, Caroline Flint, favoured a ban on cybrids. Such tensions were consistent with the broader press picture of a government at the time fragmented by a range of political issues, in contrast to the united front presented by the scientists themselves. A lack of consensus amongst ethicists on the issue was also played up, again in contrast to scientists and patient groups who were 'very clear' about the need for this type of research. As Henderson (2007c) writes, scientists showed every sign of 'borrowing the tactics of environmentalists and consumer groups to set the agenda. They patiently explained the case for carrying out controversial research to the press and the public . . .' . In the words of Henderson (2007c) scientists 'were starting to acquire the political and media savvy that they lacked in the controversies over GM crops and the MMR vaccine'.

Changing government policy

The well-managed campaign just described succeeded in reversing intended government policy. In May 2007, the government's draft legislation was republished in a revised form as a Draft Bill, which sought to ban genuine hybrids (which no scientists were actively proposing) but now proposing that the use of cybrids be permitted, as long as such work was first licensed by the HFEA. At about the same time, the HFEA instigated a new 3-month public consultation exercise to ascertain what level of public support the proposed use of cybrids attracted. When their deliberations were complete (in September 2007) the HFEA announced that scientists would be permitted to create cybrids, announcing that: 'having looked at all the evidence the authority has decided there is no fundamental reason to prevent cytoplasmic hybrid research'. This was a considerable, though not unprecedented victory for the research community, prompting one of the main protagonists to ponder ruefully on their newly found influence: 'What would have happened if the same thing had been done for GM foods . . . is a very interesting question' (quoted in Henderson 2007c).

This second consultation, orchestrated by the HFEA, proved a more accomplished and sensitive assessment of public opinion than the government's initial soundings the previous year. The opportunity for involvement was welcomed by scientists; Lyle Armstrong, who headed the second research group applying to undertake cybrid work talked of the HFEA's consultation as: 'the possibility of a further public consultation exercise gives us the opportunity to explain why the science is so very important for Britain and for humanity in general' (Anon 2007b).

The comprehensiveness of this second consultation exercise was impressive, both for its transparency and thoroughness. It involved (see HFEA 2007) a consultation document and a public dialogue strand consisting of deliberative work, an opinion poll and a public meeting; the publication that emerged included a comprehensive literature review and also reported on scientific consultation. Julian Hitchcock, senior life sciences lawyer at the law firm Mills and Reeve reported that:

The government's original proposals were based on a public consultation which researchers tended to leave to their own employing organisations, while a largely uninformed public was simply (and improbably) asked 'whether the law should permit the creation of human–animal or chimera embryos'. This complacency and failure to engage with the public led the government to its proposed ban.

quoted in Cookson (2007)

It was indeed the case that the government's initial consultation presaging the White Paper of December 2006 looked impressive enough from afar; it canvassed opinion 'extensively', receiving 535 responses, ranging from medical bodies, scientists, patient groups, ethicists and faith communities, and in those cases where hybrids was the major focus of concern, over 80% of the responses were opposed to the lifting of the proposed ban. But, as Hitchock argues, that picture was skewed in that those who did not support a ban on cybrids often contributed joint 'consensus' statements to the consultation process, while those favouring a ban offered their objections independently. When the HFEA consultation was set up, it differed from the government's preceding efforts in providing an explanation of the work involved. A very different and more nuanced conclusion was reached; for example, when lay individuals not opposed to embryo research *per se* were questioned, 65% thought that the HFEA should licence cybrids, with just 21% thinking not.

To some, the change reflected a 'remarkable turnaround in public opinion' (Cookson 2007). In the same news article, a stem cell scientist saw it as a 'lesson for all scientists that they need to be open about their work and communicate it at every level'. Fiona Fox of the SMC commented:

The HFEA consultation on public attitudes to human-animal embryos shows that when the public feel they understand the science and can see which diseases the researchers are trying to tackle, support swings strongly in favour of allowing research'

quoted in Cookson (2007)

For those scientists eager for an educative programme, following a rejuvenated 'deficit model', the outcome proved inspiring, with perhaps a return to clearer goals.

In reality, the apparent reversal of opinion raises concerns that are more intriguing and wide-ranging than such comments imply. It is tempting to believe that scientific information was the key to the change of policy described; in reality success had surely more to do with skilful political lobbying, a mobilized and seemingly united scientific community, a better understanding of public opinion and the benefit of an uneven battleground where the opposition could be portrayed as standing in the way of a healthier, happier future for humanity at large. In the words of the leading article of *The Independent* (2007)—and this from a newspaper with a sustained track record in opposing the development of GM crops:

... it is important for scientists to realise that they must never leave the public behind. The knowledge gap between the lay public and the experts may be widening, but it is up to scientists to make sure they keep us fully informed about what they are doing, and—no less important—why.

A plea for genuine public engagement with science

This chapter has been about continuity and about change; it pointed both to continuing difficulties in how controversy and consensus is communicated and to new opportunities for the communication of science. First, the disorganized unfolding of scientific understanding on issues that impact on 'real life', for example in the health sciences, create a range of seemingly contradictory guidance that often befuddle and alarm recipients—as opposed to providing the clarity that scientists fondly imagine their subject provides. Second, since conflict is an endemic and generally positive component of scientific advance, its major cheer-leaders will inevitably seek and enjoy their high public profile, just as do other celebrities, and in doing so will draw attention to science in ways that are only occasionally of benefit.

Of course, political lobbying by scientists is far from a new or unusual phenomenon, as Mulkay's (1997) study of the 'great embryo debate' in the UK in the 1980s testifies. More recently, the involvement of scientists and professional societies in the climate change debate in the UK has proved significant (see for example Ward 2007). But what is currently of special interest is the wide variety of communication tools and channels now available to scientists for that purpose, for example the different forms of dialogue, engagement and consultation fast becoming a requirement of policy-making in the UK. For good reason, scientists seem most confident when their chosen field of engagement is centred around the 'science' at the core of their case, and there is a widespread but unproven assumption that the exposition of the science will most likely persuade nonspecialists to align their views with those of scientists. It would surely be unfortunate if scientists defined their contribution to future debate so narrowly, or attributed what could be seen as public relations (PR) 'victories' to the one-way communication model that other sectors of the scientific community are assuring us is now a thing of the past.

What has been relatively cheaply and swiftly achieved in the UK is an up-and-coming generation of more 'professionalized' scientists, experienced to different degrees in 'communication and media training'. This is part of a broader emphasis amongst scientists on the purposes and methods of science communication, as the appearance of this volume shows. There is more 'water-cooler' discussion amongst scientists about how to 'work with popular and new media', how 'messages can be got across'. A plethora of formal training opportunities are available to scientists in the UK—run by the research councils and universities, for example. Increased media-savviness is coming about in a makeshift, piecemeal way but its effects are likely to be increasingly noticeable. This raising of media awareness and tactical acumen in the UK is aided by structures and systems that seek to maximize science's input—from university press officers to the UK's Science Media Centre. With the advent of dialogue, public opinion is becoming a more important and contested area of interest—and one that is seemingly all too rarely effectively researched. What is emerging is an altered communication landscape for scientists, with opportunities and expectations that take scientists into territories far removed from the laboratory bench—yet further into the uncharted ground of 'postacademic science' that John Ziman (2000) elegantly foretold. To Ziman's skills list for post-academic scientists, including commercial acumen, entrepreneurial ambition and inspired leadership can be added a comfort with journalistic practices and awareness of a range of promotional strategies.

If such developments help scientists to better understand and appreciate publics, and build on well-established interest in science, then it cannot come too soon. But the day when scientists routinely show a gift for 'getting their message across', for 'handling media', for PR and for lobbying would mark a very different culture, especially if achieved at the expense of science's more admirable reputation for openness, scepticism and ready admission of uncertainty.

■ REFERENCES

Abbott, B. (2007). Letter to the Editor: kick the carbohydrates and restore sanity to the obesity debate. *The Independent*, 9 November, p. 42.

ABC National Radio (13 March 2006). Risk factors for obesity and being overweight. *The Health Report*. Available at: http://www.abc.net.au/rn/healthreport/stories/2006/1590441.htm.

Allan, S. (2009). Making science newsworthy: exploring the conventions of science journalism. In: *Investigating Science Communication in the Information Age: Implications for Public Engagement and Popular Media* (ed. R. Holliman, E. Whitelegg, E. Scanlon, S. Smidt and J. Thomas). Oxford University Press, Oxford.

Anon (2007a). Scientists support hybrid embryos. *BBC News Online*, 10 January. Available at: http://news.bbc.co.uk/1/hi/health/6247063.stm.

Anon. (2007b). Public debate on hybrid embryos. *BBC News Online*, 11 January. Available at: http://news.bbc.co.uk/1/hi/health/6251627.stm.

Anon (2007c). Science adviser urges GM rethink. *BBC News Online*, 28 November. Available at: http://news.bbc.co.uk/1/hi/sci/tech/7113199.stm.

Bithell. C. (2007). Interview with Fiona Fox, head of the Science Media Centre. *Stempra Newsletter*. Available at: http://www.stempra.org.uk/newsletter/07_spring/03.htm.

Campbell, S. and Townsend, E. (2003). Flaws undermine results of UK biotech debate. *Nature*, **425**, 559.

Coghlan, A. (2007). Frankenbunny, human, or cybrid? *Short Sharp Science: Science News Blog from New Scientist*, 5 January. Available at: http://www.newscientist.com/blog/shortsharpscience/2007/01/frankenbunny-human-or-cybrid_05.html.

Connor, S. (2007). Medicine needs hybrid embryos, scientists say. *The Independent*, 5 April. Available at: http://www.independent.co.uk/life-style/health-and-wellbeing/health-news/medicine-needs-hybrid-embryos-scientists-say-443420.html.

Cookson, C. (2007). Lobby wins backing for hybrid embryos. *Financial Times*, 5 September. Available at: http://www.ft.com/cms/s/0/c6da6ee4–5b3b-11dc-8c32–0000779fd2ac.html.

Flegal, K., Graubard, B., Williamson, D. and Gail, M. (2007). Cause-specific excess deaths associated with underweight, overweight and obesity. *Journal of the American Medical Association*, **298**(17), 2028–37.

Fleming, N. (2007). MPs oppose animal-human embryo ban. *The Telegraph*, 6 April. Available at: http://www.telegraph.co.uk/news/main.jhtml?xml=/news/2007/04/05/nembryo05.xml.

GM Nation? (2003). *GM Nation? The findings of the public debate*. Available at: http://www.aebc.gov.uk/aebc/reports/gm_nation_report_final.pdf.

Henderson, M. (2007a). 'Cybrid' exemption for stem-cell research. *The Times*, 1 March. Available at: http://www.timesonline.co.uk/tol/news/uk/science/article1454609.ece.

Henderson, M. (2007b). Blair faces revolt over law to ban 'chimera' research. *The Times*, 5 April. Available at: http://www.timesonline.co.uk/tol/news/uk/science/article1615317.ece.

Henderson, M. (2007c). White coats defeat the grey suits in a clash between science and Whitehall. *The Times*, 13 October. Available at: http://www.timesonline.co.uk/tol/news/uk/science/article2648626.ece.

Herold, E. (2006). *Stem Cell Wars; Inside Stories From the Frontline*. Palgrave Macmillan, New York.

Horlock-Jones, T., Walls, J., Rowe, G., Pidgeon, N., Poortinga, W. and O'Riordan, T. (2004). *A Deliberative Future? An Independent Evaluation of the* GM Nation? *Public Debate About the Possible Commercialisation of Transgenic Crops in Britain, 2003*. Available at: http://www.cf.ac.uk/socsi/research/linksresources/gm_future_top_copy_12_feb_04.pdf.

House of Commons, Select Committee on Science and Technology (2007). *Government Proposals for the Regulation of Hybrid and Chimera Embryos*, Fifth Report. HMSO, London.

House of Lords, Select Committee on Science and Technology (2000). *Science and Society*, Third Report. HMSO, London.

HFEA (Human Fertilisation and Embryology Authority) (2007, April). *Hybrids and Chimeras. A Consultation on the Ethical and Social Implications of Creating Human/Animal Embryos in Research*. Available at: http://www.hfea.gov.uk/en/1511.html#hybrids.

Human Genetics Alert (HGA) (2007). Memorandum 51: submission from Human Genetics Alert to the House of Commons Select Committee on Science and Technology. *Government Proposals for the Regulation of Hybrid and Chimera Embryos*, Fifth Report, pp. Ev 132–137. HMSO, London. Available at: http://www.publications.parliament.uk/pa/cm200607/cmselect/cmsctech/272/272ii.pdf.

Hunt-Grubbe, C. (2007). The elementary DNA of Dr Watson. *Sunday Times*, 14 October. Available at: http://entertainment.timesonline.co.uk/tol/arts_and_entertainment/books/article2630748.ece.

Irwin, A. (2009). Moving forwards or in circles? Science communication and scientific governance in an age of innovation. In: *Investigating Science Communication in the Information Age: Implications for Public Engagement and Popular Media* (ed. R. Holliman, E. Whitelegg, E. Scanlon, S. Smidt and J. Thomas). Oxford University Press, Oxford.

The Independent (2007). The arguments for hybrids. Leading article, 6 September. Available at: http://www.independent.co.uk/opinion/leading-articles/leading-article-the-argument-for-hybrids-401493.html.

Jasanoff. S. (1997). Civilization and madness; the great BSE scare of 1996. *Public Understanding of Science*, 6, 221–32.

Kohn, M. (2007). Science, snakes and ladders. *The Independent*, 2 November, p. 25.

Latour, B. (1987). *Science in Action*. Open University Press, Milton Keynes.

Macrae, F. (2007). Hybrid embryos could be created within months. *The Daily Mail*, 6 September. Available at: http://www.dailymail.co.uk/pages/live/articles/news/news.html?in_article_id=480039&in_page_id=1770.

Martin, B. and Richards, E. (1994). Scientific knowledge, controversy and public decision making. In: *Handbook of Science and Technology Studies* (ed. S. Jasanoff, G. Markle, J. Petersen and T. Pinch), pp. 506–26. Sage, London.

McKie, R. (2007). Disgrace; how a giant of science was brought low. *The Observer*, 21 October, pp. 28–9.

Minger, S. (2007). Engaging through the media and PR. *Report of the BA Science Communication Conference*. Available at: http://www.the-ba.net/the-ba/ScienceinSociety/ScienceCommunicationConference/_2007_Conference_Report.htm.

MORI (2001). *The Role of Scientists in Public Debate*. Research Study conducted by MORI for the Wellcome Trust. Available at: http://www.mori.com/polls/2000/pdf/wellcome-main.pdf.

Mulkay, M. (1997). *The Embryo Research Debate; Science and the Politics of Reproduction.* Cambridge University Press, Cambridge.

Nelkin, D. (1994). Scientific controversies; the dynamics of public disputes in the United States. In: *Handbook of Science and Technology Studies* (ed. S. Jasanoff, G. Markle, J. Petersen and T. Pinch), pp. 444–56. Sage, London.

Nowotny, H., Gibbons, M. and Scott, P. (2001). *Rethinking Science: Knowledge and the Public in an Age of Uncertainty.* Polity Press, Cambridge.

Office of Science and Innovation (2006). *The Government's Approach to Public Dialogue on Science and Technology.* Available at: **http://www.sciencewise.org.uk/html/sitedocuments/ 1Sep06GuidingPrinciplesforPublicDialogue.pdf**.

Porter, H. (2007). His views are hateful; but so is the attempt to deny him a voice. *The Observer,* 21 October, p. 31.

Roberts, M. (2007). Why are ministers opposed to hybrids? *BBC News Online,* 5 January. Available at: **http://news.bbc.co.uk/1/low/health/6233671.stm**.

Rose, S. (2007). Watson's bad science. *The Guardian: Comment is Free . . .* Available at: **http://commentisfree.guardian.co.uk/steven_rose/2007/10/watsons_bad_science.html**.

Science Media Centre (2007). Scientists react to Department of Health: Human Tissues and Embryos (Draft) Bill. *Press Release,* 17 May. Available at: **http://www.sciencemediacentre.org/ press_releases/07-05-17_fertilitybill.htm**.

Stilgoe, J. and Wilsdon, J. (2009). The new politics of public engagement with science? In: *Investigating Science Communication in the Information Age: Implications for Public Engagement and Popular Media* (ed. R. Holliman, E. Whitelegg, E. Scanlon, S. Smidt and J. Thomas). Oxford University Press, Oxford.

Thomas, J. (1997). Informed ambivalence. In: *Science Today; Problem or Crisis?* (ed. R. Levinson and J. Thomas), pp. 163–72. Routledge, London.

Thomas, J. (1999). A genetic basis for public enlightenment? A personal view of the Dawkins/Pinker phenomenon. *Public Understanding of Science,* **8**, 345–51.

Topping, A. (2007). Professor who created Dolly the sheep to abandon cloning. *The Guardian,* 17 November. Available at: **http://www.guardian.co.uk/science/2007/nov/17/stemcells**.

Usbourne, D. (2007). Now doctors say it's good to be fat. *The Independent,* 8 November, pp. 1–2.

Ward, R. (2007). The Royal Society and the debate on climate change. In: *Journalism, Science and Society: Science Communication Between News and Public Relations* (ed. M. Bauer and M. Bucchi), pp. 159–72. Routledge, London.

Washington Times (2005). Gobble, gobble. Editorial, November. Available at: **http://findarticles.com/p/articles/mi_hb5244/is_200511/ai_n19624925**.

Watson, J. (1968). *The Double Helix.* Weidenfeld and Nicolson, London.

Watson, J. (2007). *Avoid Boring People; and Other Lessons From a Life in Science.* Oxford University Press, Oxford.

Ziman, J. (2000). *Real Science.* Cambridge University Press, Cambridge.

■ FURTHER READING

- Collins, H. and Pinch, T. (1998). *The Golem; What Everyone Should Know About Science,* 2nd edn. Cambridge University Press, Cambridge. This book is about what makes science and scientists tick, and one that reveals the untidiness of science, as opposed to formal and sanitized narratives. Controversies in science feature large, with the authors arguing that they indicate the true nature of scientific processes.

- Brante, T., Fuller, S. and Lynch, W. (1993). *Controversial Science; From Content to Contention.* SUNY Press, Albany, NY. This is a richly diverse and engaging collection of articles that cover a range of theoretical and culturally important aspects of controversies in science. Sociological perspectives are much to the fore, as are historical and ethical aspects, painting a picture still relevant to today's context.

- Levinson, R. and Thomas, J. (1997). *Science Today; Problem or Crisis?* Routledge, London. This book has important things to say about the communication of controversy, particularly in relation to the formal and informal learning of science and to why teaching controversies to students is both necessary and problematic.

■ USEFUL WEB SITES

- **Science Media Centre (SMC): http://www.sciencemediacentre.org/**. This UK-based organization is at the forefront of attempts to raise the profile of science in news media. A major role of the SMC is to help journalists link up with scientists of appropriate expertise. Also evident here is information and resources that relate to broader roles of lobbying and briefings, reflecting what is a more pro-active and assertive role for this non-governmental organization.

- **Human Fertilisation and Embryology Authority (HFEA): http://www.hfea.gov.uk/**. This licensing organization has its critics, but its web site is an impressive example of accessible and non-partisan information. The full range of the HFEA's functions is reflected in diverse resources, including reports of past controversial issues associated with reproductive technologies, many involving public consultation exercises.

- **The Wellcome Trust: http://www.wellcome.ac.uk/**. This medical charity supports research to improve human (and animal) health but its public engagement programme and funding roles are also worthy of note. Along with other charities, the Wellcome Trust is active in promoting science, so disseminating research findings on key health issues is an important priority, for example by promoting conferences.

- **Guardian.co.uk; controversies in science: http://www.guardian.co.uk/science/controversiesinscience**. The prevalence of science-based controversies in news coverage means that this blog is a fascinating repository of newsworthy topics, written about in an engaging way. On most recent access (February 2008) no fewer than 82 controversial topics were under discussion.

SECTION 5
Popularizing science

The reader, at this point, will have realised for some time that this is not a chemical treatise: my presumption does not reach so far. . . . Nor is it an autobiography, save in the partial and symbolic limits in which every piece of writing is autobiographical, indeed every human work; but it is in some fashion a history.

Primo Levi (1975) *The Periodic Table*

5.1   **Where do books fit in the information age?,**
      by Bruce Lewenstein   151

5.2   **Science communication in fiction,** *by Jon Turney*   166

5.3   **Speaking to the world: radio and other audio,**
      by Martin Redfern   178

5.1

Where do books fit in the information age?

Bruce V. Lewenstein

Are science books important?

Why should we care about science books? After all, we live in a 'new media' world where students, researchers and members of the public use the World Wide Web for so many of their information needs. Cutting-edge research appears on 'preprint archives' or 'open-access' online journals (see Chalmers, Chapter 2.2 this volume; Gartner, Chapter 3.2 this volume), 'textbooks' appear as online sites with interactive presentations, with links to public discussion and dialogue, and even for archiving current research. In that kind of world, what's the purpose of looking at 'old fashioned' books?

In fact, I want to argue, books are tremendously important in science. They provide structure and substance for scientific communities—both communities within scientific practice and communities of scientific interest that extend beyond the professional scientific world, communities that encompass various publics and define their inter-action with science. Science books can be understood as shared social experiences, ones that through their use create a common bond that may or may not be based on the actual content of the text. In some cases, the books may serve multiple communities, crossing boundaries in complex ways. Books serve as social memories, providing cultural touch-points that allow communities to express their common norms and interests.

To explore these issues, I will look at books in several categories: books of daily use such as reference books, textbooks, those with clear influence on intellectual culture and those with clear influence on broader public culture—what the French call '*culture scientifique*', or the place of science and scientific ideas in the cultural matrix. I will look at how scientific ideas are presented, conveyed and used to create intellectual regimes, as well as how they are used in discourses that both contribute to science's social authority and simultaneously allow the ideas to shift meanings as they get used in different contexts.

While many of my examples will be drawn from the United States in the years after World War II (post-1945; the period I know best), the argument extends throughout the developed world: science books circulate internationally and many publishers prominent in American science are outposts of European publishers. Moreover, the definition of 'book' is itself flexible; as the digital world of 'blooks' (intersections of blogs and books), e-books and portable digital reading devices expands, the capabilities and contexts of books and book use are changing rapidly.

Books 'within' science

Books in the daily life of science

Many scientists will say that 'if it doesn't appear in a journal, then it's not science'. But at least through the end of the 20th century, books clearly retained a place in the daily practices of science. Those daily practices show how books create a sense of community. Major examples include reference books like the *CRC Handbooks* (of math, chemistry, and so on) or the data base serials that until recently were bound and treated as books (such as *Chemical Abstracts* or *Science Citation Index*). Reference books represent a form of standardized knowledge, and their widespread use implies a communal judgement about which standards to use and which references to rely upon. These judgements are not merely matters of convenience, but also clear statements about trust and the establishment of networks of interaction that create the social fabric through which scientific development is woven. Other reference books contain the consensus from those networks about the findings of recent research, such as the volumes produced as *Annual Reviews* (of biochemistry, of physiology, of energy and the environment, and so on) (Kaufmann 1995).

Yet reference books also provide a marker of changes in the use of books across the post-1945 period. Saying the name *CRC Handbook* out loud in a group of scientists trained before the 1990s leads to a collective sigh of recognition that demonstrates their communal power to a particular generation. More recent generations turn not to hardcover books but to electronic data bases, some of which may be identified more with the web site at which they are housed than the 'book' in which they appeared. The individual articles in *Annual Reviews*, for example, are available as free-standing document files; the publisher of *Annual Reviews* worries that readers will no longer identify with the underlying books and therefore will be unwilling to convince their libraries to subscribe (S. Gubins, personal communication, 31 January 2002). The tables and constants for which earlier generations turned to the *CRC Handbooks* are now available on various web sites and calculators. The book is no longer the repository of stored knowledge to working scientists; web sites available through the internet serve that function instead.

These new electronic forms also change the daily practices of science. The 'semantic web' (new electronic standards that allow automated linking of online resources) means that electronic preprints in astronomy may link directly to reference data bases (which used to be books) of astronomical objects which link directly to previously published journal articles, with all the information displayed in dynamic pages that highlight congruencies, discrepancies and missing information—all of which create new connections between researchers and ideas, generating new knowledge in new ways (Henneken *et al.* 2007). 'Books', in their traditional form, do not appear in these practices, yet the habits of collation, referencing and reading that books engender are central to the production of new knowledge.

Those habits are fundamental to the processes by which knowledge and communities are created. For at the same time that working scientists have reduced their dependence on books as a source of stored knowledge, books have continued in that role among broader audiences. The lowered real costs of books and improved distribution systems created as a result of broad changes in the publishing industry have made scientific reference texts much more widely available. This has created networks of people using scientific informa-tion in their daily lives that extend beyond research scientists, including members of the general public with no professional need for technical information. Recent editions of the *Merck Manual*, a compendium of medical information, for example, have sold about a million copies in the United States alone, with both authorized and pirated editions published world-wide (Berkow 1987).

Another form of books used in the daily life of scientists are conference proceedings. Proceedings are again evidence of community, since they are literally documentation of communal efforts, of occasions when scientists came together to work through their ideas. In 1965, the *Philosophical Transactions of the Royal Society* published the papers presented at the first symposium to systematically bring together the evidence for what was then called 'continental drift' (Blackett 1965). Though the *PhilTrans* (as aficionados call it) is technically a journal, its compilation of papers from a single event in a single volume created a book that became a classic. Today, in fields such as computer science, conference proceedings are often the primary form of formal publication. The proceed-ings are now often online, yet the concept of the book—the single, cohesive unit defined by the connection to a particular meeting or conference—is the intellectual idea that binds the work together.

Similarly, the communal event documented in a *festschrift* both celebrates an individual and collates an intellectual programme. A *festschrift* is a volume presented to a senior researcher at a milestone such as retirement or a 60th or 75th birthday. It often includes papers by former students and colleagues reporting on topics from the recipient's career, bringing them up-to-date and highlighting the ongoing trajectory of the work. The nature of the *festschrift* is to highlight the personal bonds that give shape and substance to scientific communities. Production of a *festschrift* is a statement about shared values, a commitment to science as a community as well as a body of knowledge. *Festschriften* can also contain important science. The most well-known example may be a paper by geophysicist Harry Hess which became one of the founding documents of plate tectonics. Because of the speculative nature of Hess's ideas (in the paper, he called them 'geopoetry'), he had trouble publishing the paper in regular journals; a *festschrift* provided the necessary outlet (Hess 1962).

Despite the value of *festschriften*, conference proceedings, technical reports and other elements of the daily life of science, many of them had a somewhat ephemeral existence in the years after World War II. They existed in a realm of publishing difficult to access reliably: the world of 'grey literature', issued by organizations without the formal apparatus of publication records or library or retail distribution. During the 1960s and 1970s, these documents were often photo-offset printed from typewritten manuscripts, distributed in idiosyncratic ways by organizations or meetings without systematic publishing operations. The growth of the World Wide Web has changed the nature of grey literature: now often posted on institutional or organizational web sites, these documents of the daily life of communal scientific work are now accessible through both general search engines such as Google[1] and more specialized sites tailored to specific scientific communities. At one level, this makes the literature—and thus the communal activity—more broadly available. At another level, accessibility through the internet is even more ephemeral than traditional grey literature, with a substantial fraction of all web sites disappearing over the course of even just a few months (Henderson 1997).

Textbooks

While the books of daily practice document science as a community, textbooks show even more clearly how books can create and shape a community. In the early 19th century, for example, a concern for bringing young women into the educated classes led to Mrs. Marcet's *Conversations on Chemistry* (initially published in London and then in a great many editions in both England and the United States), one of the most successful textbooks of the time (Crellin 1979; Lindee 1991; Dolan 2000; Knight 2000). The historian Jon Topham has shown how the development of analytical mathematics at Cambridge in the early 19th century, normally described as something limited to a small set of leading mathematicians, was strengthened by its extension to a much larger student community through the efforts of publishers seeking to address the needs of a local market (Topham 1998). A hundred years after its first publication, Hardy's *A Course of Pure Mathematics* remains a bestseller.

More recently, access to higher education has increased and textbooks sell tens of thousands of copies when adopted for use in large introductory courses in colleges and universities. In the United States, Pauling's *General Chemistry* (1947), Sears and Zemansky's *College Physics* (1947), and Morrison and Boyd's *Organic Chemistry* (1959) shaped knowledge in their fields for years to come. Leading introductory physics textbooks, such as Sears and Zemansky's, structured their introductions to physics around the needs of engineering students. While they added principles and abstraction to earlier texts (which had often been organized around specific experiments or demonstrations), they were fundamentally interested in teaching the Newtonian mechanics that most physics students would have to face. As the physicist and historian Charles Holbrow has shown, quantum mechanics and other issues of 20th century physics were almost literally an appendage, appearing in separate chapters near the ends of these texts (Holbrow 1999). For many work-a-day physicists, not the elite researchers but the ones keeping government laboratories and experimental facilities running, the ones doing routine calculations and operations, the engineers who make up the vast bulk of the physically oriented scientifically trained workforce, esoteric cutting-edge science was

1. http://www.google.com.

just one part of what they learned. The world-view, the intellectual matrix into which they placed the individual facts, theories, formulae and behaviours that to them *defined* science, that world-view was essentially a 19th century world-view. For most purposes, the older world-view is productive: in the every-day case, non-quantum approaches appropriately approximate the natural world. But the use of textbooks that stressed the older approaches means that for most scientists trained in 'modern physics', quantum issues are not foremost in their minds.

Textbooks, especially at the advanced level, can also be the place for creating new fields. James Watson's *Molecular Biology of the Gene* (1965) was intended both as a text and as an intellectual argument for a new field. Watson's more famous book, *The Double Helix* (1968), has been interpreted as a polemic arguing for a new, competitive, high-stakes approach to biological research (Yoxen 1985). Less recognized has been the role of the earlier text in creating a new discipline. Watson brought together the range of research that had previously been scattered in crystallography, biochemistry, genetics and other fields to show that it could be taught together fruitfully—and that, by so doing, teachers could train a new generation of scientists ready to fully inhabit this coherent area rather than merely reaching into it from their own home disciplines (Watson 1965; Brenner 2000). Similarly, E. O. Wilson's *Sociobiology* (1975) was an explicit argument for a new approach to evolutionary research. Paul Samuelson's *Economics*, especially in its first two editions, established the authority of rational choice theory as the leading model in economic thought (Samuelson 1948).

The power of textbooks is their ability to create communities of people with similar training, similar perspectives and similar tools. Because (unlike most countries) the United States does not have a national curriculum, the fragmentation of science training is a particular concern there. In the 1990s, as efforts gathered steam to reform the American science curriculum in public elementary and secondary schools, much attention was put on the ability of textbooks not only to convey information, but also to convey an overall sense of what science was and how it should be related to public life (Finley and Pocovi 1999).

Books in intellectual and public culture

There are many ways to identify books that play a role in intellectual and public culture. Some have a kind of official presence: they have won an award, a Booker Prize, a Pulitzer Prize, a National Book Award, etc. Others have become bestsellers. Yet others fall into a category I call 'remembered books', the ones where someone I'm talking with remembers the book and then says 'But you're going to include *that* book aren't you?'. These are the books that have become touchstones for intellectual culture.

Perhaps the most famous book linking science and intellectual debate is C. P. Snow's *The Two Cultures* (first published in 1959 after being delivered as the BBC's Rede Lecture). Snow argued that the cultures of art and science were diverging, but that both were needed for addressing the pressing needs of a rapidly developing world. Reaction to his lecture was muted at first, but several years later the prominent literary critic F. R. Leavis challenged Snow in a series of essays (Snow 1959, 1963, Leavis 1963). The particular disputes between Snow and Leavis may have had as much to do with personal animus and institutional competition as with the merits of their arguments (Boytinck 1980; Ortolano 2002). But the continuing prominence of the debate about 'two cultures', even 50 years after its introduction, speaks to the ability of books based on science to become part of general discourse.

Other books frequently 'remembered', at least among the scientific community, include Paul de Kruif's *Microbe Hunters* (1926) and Sinclair Lewis's *Arrowsmith* (1925). In Britain, the group of scientists gathered into the 'social relations of science' movement in the years between the world wars (1918–1939) produced a series of books often cited as models of explanation and integration of science and culture: among them, Lancelot Hogben's *Mathematics for the Million* (1936) and *Science for the Citizen* (1938), Julian Huxley's *Scientific Research and Social Needs* (1934), and especially the crystallographer J. D. Bernal's *The Social Function of Science* (1939). Each of these books took on the belief on the part of scientists that they had a contribution to make that was fundamental to the construction of modern culture (Werskey 1971, 1978).

Yet in the United States, in the first 30 years after World War II, few science books won Pulitzer Prizes (the most prominent award there for books). One book, James Phinney Baxter's *Scientists Against Time* (1948), published right after the war, was a story about the atomic bomb. William Goetzmann's *Exploration and Empire* (1967) was about exploration of the American West, and Rene Dubos's *So Human an Animal* (1969) drew heavily on his background in biology. Then the pattern changed. Beginning with Carl Sagan's *Dragons of Eden* in 1978, then every year or every other year into the late 1990s, the Pulitzers honoured a science book. These books appeared in both the general non-fiction and the history category of the Pulitzers. Clearly something happened in the late 1970s to make science books more central to American culture (I don't yet know the story sufficiently well in other countries to make similar claims elsewhere). Science became a part of the general intellectual discussion. Interestingly, that same time period is also about the time of a 'science boom' in popular science magazines, television shows and science museums (Lewenstein 1987). The relationship between science and American culture went through a change in the late 1970s, such that science became a necessary part of any cultural discussion.

A similar pattern appears in American bestsellers. Before the mid-1970s, only rarely did more than 10 new science-oriented books a year become added to the list of bestsellers maintained by the *New York Times*. But after 1978, only rarely did *fewer* than 10 science-oriented books get added to the list. More science books were being sold. That's another marker to suggest that science had become a necessary part of ongoing cultural conversations. The Pulitzer Prize data and the bestseller data suggest that, despite the ongoing reference to 'two cultures', the idea that those cultures don't speak to each other may no longer hold (if it ever did).

To understand this new cultural debate, we need to know more about what specific types of books were appearing on the bestseller lists. There are at least two kinds. First are the books in which 'science' appears as a main character. These are the books that are about physics, or astronomy, or biology or so forth (Mellor 2003). The second set of books are those that I call 'public science'. These books are about, for example, sex, but they draw on the science of sex. These are the inspirational books that draw on psychological research. Many of the diet, health, fitness and medicine books draw on scientific research or at least the appearance of scientific research. Even if some of these books don't use science well, they get some of their credibility precisely because they lay claim to the authority of science. Some people argue that science is not valued in our society. I disagree. These books become bestsellers by claiming to draw on science, which they do because science is respected in the community of ideas. The book data indicate that science actually plays a very important and respected role in general culture.

Another type of book is the 'grand' books, such as Jacob Bronowski's *Ascent of Man* (1973) or Carl Sagan's *Cosmos* (1980). Following in the tradition of the Hogben and Bernal books, these broad sweeps of scientific ideas become bestsellers only in the 1970s. The breakthrough in the United States was most clear in 1980 with Sagan's *Cosmos*. As with the Bronowski book, it was tied to an extremely popular television show, and that helped drive the sales. But the book itself was also a bestseller—a bestseller so great that shortly after it was published, Sagan was given a $2 million contract for what would become the novel *Contact*. At the time, that was the largest advance ever given for a fiction book that was not even in manuscript form. *Cosmos* marked the moment that something different was clearly going on.

In the 'science as science' category, the next big moment was Stephen Hawking's *A Brief History of Time* (1988) (Maddox 1988; Rodgers 1992). Hawking's book is the one that everybody bought but nobody read. He said in the introduction that he left out all the mathematical equations so that he wouldn't lose readers, but the book is still pretty tough to read. In Britain, the book sold nearly half a million copies in its first few years. In the United States, it sold 700,000 copies in hardcover in its first year and 400,000 copies in its second year. By 1992, the book had sold 4.5 million copies world-wide. Even today, 20 years later, new editions continue to appear. It set a new sort of expectation about what books can accomplish. Hawking's book opened up the book publishing world—and thus the broader cultural world—to science. After it appeared, science books began to receive aisles in American bookstores, agents went seeking authors to write books about engaging in science (Romano 1995; Barringer 1998).

All of this evidence suggests that books have played a role in general culture, certainly in the United States. Some of the evidence shows that books have become even more important after the mid-1970s than they were shortly after World War II, even as alternative media such as television and eventually the internet have become more popular. As the boundaries between traditional books, digital media and new forms of science such as blogs and blooks become fuzzier, the importance of 'books' in science will likely remain high.

How are science books important?

Books exercise their cultural importance by contributing to public discussion in four areas.

First, books are important for the intellectual development of science itself. Even though some of the bestselling or prize-winning books are targeted to the public, they are also targeted to the scientific community or they play a role within the scientific community. That should not surprise us, given current conceptual understandings that science communication involves feedback among different forms of communication and loops that connect different types of communication (Lewenstein 1995).

The second role that books play is to recruit people into science. By making science exciting and accessible, books help young people imagine themselves in jobs and activities that they haven't yet personally experienced. As noted above, scientists have often cited books by de Kruif, Hogben and Watson as their inspiration for becoming researchers.

The third role is one that cannot easily be expressed in English. The French call it *culture scientifique*, the idea of everyday culture as infused with science. If we say 'a scientific culture' in English, it doesn't carry the same meaning that it seems to carry in French-speaking countries. The idea is that books show the integration of science and culture in our everyday life.

The final role is one of public debate, in which books are the location or the forum in which public issues can be discussed.

Intellectual development of science itself

For an example of how a prize-winning book contributes to science itself, consider E. O. Wilson's *Sociobiology* (1975). This book was partly intended for the 'science-attentive public', the elite intellectual community, the 'chattering classes'. But it was also an argument within science itself. It was Wilson's full, complete statement of the sociobiology programme. It was intended for use within the scientific community as a statement of that programme. In a very real sense, it pulled that field together, making explicit some of the connections and ideas that had previously existed only in separate papers or only in specialist communities. Wilson's book made the new field concrete (Segerstråle, 2000). Richard Dawkins's *Selfish Gene* (1976) also contributed to the creation of sociobiology, leading in 2006 to a *festschrift* entitled *Richard Dawkins: How a Scientist Changed How We Think*.

A similar function was played by one of the textbooks listed above, James Watson's *Molecular Biology of the Gene*. That book pulled together the field of molecular biology, which had not existed before. Whole courses were created to teach that textbook. In the same way, courses were suddenly created called 'Sociobiology', based on Wilson's book, pulling together the field in a way that had not been in evidence before. Yet, especially because of Wilson's last chapter on humans, the book also became part of a general public discussion about the nature of who we are.

Another example is Joseph Weizenbaum's *Computer Power and Human Reason* (1976). The book is a key text within artificial intelligence. At the same time, it is also part of the general discussion about the role of computers in society, the workings of the human mind and all those related topics.

James Gleick's *Chaos* (1987) is interesting because it also seems to serve this intellectual role within science, even though it was written as a popular science book. Gleick was just another journalist going out and writing a book that would explain some area of science. Yet the book served the function of pulling together the field of complexity and chaos in a new way. More recent books in the field cite Gleick's book as one of the things that pulled people together, that made them suddenly realize that they were all talking to each other (Lewin 1992; Waldrop 1992). The public discussion shaped the intellectual discussion as well—through the medium of books.

Recruitment

Recruitment books pull people into science. These are books that people cite as 'Hey, the reason I'm a scientist is because I read that book'. De Kruif's *Microbe Hunters* is the epitome of these books. It is astonishing how frequently that book appears in the memories (and sometimes memoirs) of senior scientists who became biologists in the 1930s, 1940s and 1950s—they read *Microbe Hunters* and that's what turned them on (Henig 2002). Similarly, Hogben's *Mathematics for the Million* drew in young scientists fascinated by numbers, precision and the possibilities of mathematics.

James Watson's *Double Helix* is a very different kind of book, but served much the same purpose in the 1960s, 1970s and maybe even the 1980s. People who in the late 20th century were at the forefront of biotechnology or genomics read Watson's book as graduate students and said 'Yeah, that is the kind of scientist I want to be! I get to make a Nobel prize-winning discovery, and then I get to go play tennis, and then I get to go get the girls'. That sounded like a fun kind of career (Yoxen 1985; McElheny 2003).

More recently, particularly in astronomy or physics, people say that *Cosmos* (either the television show or the book) served the same function. These are often those who were so turned on by the television show that they went out and got the book. *Cosmos* has had the same kind of recruiting power as the De Kruif and Watson books: 'Why are you an astrophysicist or an astronomer?' 'Because I saw Carl Sagan's *Cosmos*' or 'I read *Cosmos*'. His popularity led to two biographies published quickly after his death in 1996 (Davidson 1999; Poundstone 1999). More than 10 years after his death, a Google search for 'influence of Sagan's Cosmos' yields blogathons, memorial web sites, tributes and other evidence of his continuing influence.

Culture scientifique

The third role of books is to create a *culture scientifique* (Fayard 1988). The essential complaint of C. P. Snow was the lack of a *culture scientifique*. But by the end of the 20th century, some aspects of such a culture seemed to exist, including the expectation among 'the cultured' that one had read some science books. Darwin's *Origin of Species* (1859) and the works of Freud and Jung fall into this category. Primo Levi's *The Periodic Table* (1975), once named the 'best science book ever', falls into this category, as well (Turney 2006). Books by Isaac Asimov and Stephen Jay Gould are 'required reading' in cultured circles (although the list does change over time—Asimov is probably less read now than he was during his lifetime). People can't consider themselves as cultured if they haven't read the essays of Lewis Thomas about medicine, Oliver Sacks about the mind or Dava Sobel's *Longitude* (1995). Not all of these books have tremendous amounts of 'science' in them—Thomas's essays are as much about philosophical approaches to illness as they are explanation of disease, and Sobel's book is more adventure story than science explanation. But one is 'expected' (in some circles) to have read those books. Among the 'science attentive' public, one is expected to have seen the excerpts of these sorts of books in the *New Yorker* or the essays on them in the *Times Literary Supplement*.

With the arrival of television, popular media began to interact to create a *culture scientifique*. Many popular science books of the 1940s and 1950s began as radio addresses.

Jacques Cousteau's first book, *A Silent World* (1953), was followed by a film with the same title (co-directed by Louis Malle) 3 years later. Bronowski's *Ascent of Man* and Sagan's *Cosmos* were both accompanied by bestselling books that substantially extended the intellectual heft of the television series, creating an expectation that one could discuss the topics raised by the shows. The combined books and films of David Attenborough (*Life on Earth, The Living Planet* and many more) fill this niche, as well.

The place of digital media in *culture scientifique* is not yet clear. Evidence suggests that many people turn to the internet for information about science and technology, and as supplements to television shows about science (Horrigan 2006; Madden and Fox 2006). But online venues seem not to be the place for extended intellectual explorations of science-related topics. One exception may be edge.org,[2] a site developed and produced by the impresario John Brockman, who came to prominence as the agent who could sell science books to publishers. Brockman himself has now edited a series of books drawing on the pronouncements and observations of the people he publishes on edge.org, attempting to claim a 'third culture' that transcends Snow's dichotomy (Brockman 1995, 2002, 2006, 2007).

Brockman's use of the web does highlight a new feature of *culture scientifique*. Many authors in all fields, but especially those of books about science that are trying to contribute to broader issues, now combine a blog with their books. Sometimes they use the blog specifically to market a book, other times they use the blogs to keep a presence in the reader's mind between books. The publishers of *Seed Magazine* (which has been described as a '*Vogue*' for 'image-conscious science aficionados') produce a collection of blogs at scienceblogs.com. Many of the authors there, such as Chris Mooney[3] (author of *The Republican War on Science* and *Storm Warning*) and Carl Zimmer[4] (author of *Soul Made Flesh, Parasite Rex* and *Evolution: the Triumph of an Idea*), use their blogs to promote their books.

Public debate

The final role is the role of public debate or public opinion. Books do not just provide information, nor do they just excite people. Some of them are in fact making arguments. Rachel Carson's *Silent Spring* is the most obvious example. That book made an argument about chemicals in society, and is widely cited as being the founding document of the environmental movement. The argument did not go uncontested. Carson's book was not attacked just by chemical companies, it was attacked by science writers. In 1963, a well-known science writer named Lawrence Lessing won the American Chemical Society's Grady-Stack Award (for excellence in science journalism). As part of his award speech, he called Carson's book 'highly emotional with a biased thesis' (Lessing 1963). Much of his talk was an attack on *Silent Spring*. This example demonstrates the degree to which there was an argument which many people felt they needed to take up.

2. http://edge.org/.

3. http://www.scienceblogs.com/intersection/.

4. http://www.scienceblogs.com/loom/.

Similarly, Evelyn Fox Keller's *The Feeling for the Organism*, a biography of Barbara McClintock, was part of the late 20th century discussion about the nature of science and whether feminine science was somehow different from masculine science (Comfort 2001). Did McClintock do science differently? Did she have some kind of female connection with her materials that males didn't have? Fox Keller, her readers, and even McClintock herself disagreed on the answers to these questions, but the important point is that Fox Keller was engaging in public discussion of a contemporary issue. Richard Herrnstein and Charles Murray's 1994 *The Bell Curve* (which argued that racial differences can explain some differences in intelligence) is similar: many people argued with the science in it, they argued about whether it properly reported research findings or interpreted data correctly (Fraser 1995; Fischer 1996; Kincheloe *et al.* 1996; Lemann 1997). But the book became a topic of discussion, with public debates, opinion stories, magazine and newspaper articles that cited it, policy discussions, etc. As with *Silent Spring* and Dawkins's *Selfish Gene*, *The Bell Curve* showed how books can play a role in public discussion.

Another function of these books, as with the blogs described earlier, is to provide the authors with a public platform. Watson's many books kept him in the public eye long after his own scientific work had given way to administrative duties. Indeed, it was during the 2007 tour for *Avoid Boring People: Lessons from a Life in Science* that Watson said that he was 'inherently gloomy about the prospect of Africa' as 'all our social policies are based on the fact that their intelligence is the same as ours—whereas all the testing says not really' (Hunt-Grubb 2007). This statement, so similar to the position presented by Herrnstein and Murray, demonstrated the ongoing power of the debate about the link between genetics and intelligence. Public furore over Watson's remarks led him, at the age of 78, to resign his final official positions and retreat (at least temporarily) to a fully private life (Dean 2007; see Thomas, Chapter 4.2 this volume for further discussion).

Watson is not alone in courting controversy with his books. Richard Dawkins, for example, has frequently published books on public issues; in 1995, he became Professor of Public Understanding of Science at Oxford. In the early 2000s, he began to focus on what he believed to be the pernicious effects of religious belief. His book *The God Delusion* (2006) was one of a series of books published around the same time that did more than question the existence of God—they actively argued against the possibility of God's existence (one title is *God: The Failed Hypothesis*) (Dennett 2006; Harris 2006; Stenger 2007).

All these examples demonstrate the continuing power of books to provide a space, a location, for discussion of issues drawing on science but intersecting with other issues in public life.

Conclusion

Books drive public discussion, most simply because they are part of the media mix that permeates our culture. While we focus on the World Wide Web and other new media because of their freshness, we can't forget that there are lots of other components of science communication; books are there. More deeply, books drive public discussion because of the multiple roles they play in providing information, engaging different forms of expertise and contributing to public discussion.

Books bring new perspectives into science. As we think about the functions of public communication of science and technology, we need to remember examples like *Chaos*, the book in which the journalist James Gleick pulled together an intellectual field in a way that hadn't been done before. We need to think about the stimulating of discussion —not just making a reader feel good the way a Lewis Thomas book did, but making the reader argue with a book in the way the Herrnstein and Murray book did. That is a role that books can play. That role highlights the place of books in public participation models of science communication.

Ultimately, books help to create the culture that we live in. They are elements both of the scientific culture and of our more general culture. By looking at them we can actually see the ways in which science and modern culture are not separate but are interwoven. Neither science nor society exists without the other one. Books provide an example of how that interaction exists in a real, material way. If we think about the multiple ways that books demonstrate the interaction of science and society, then we can begin to understand their enduring importance.

Acknowledgement

Substantial parts of this chapter have been previously published as Lewenstein, B.V. (2007). Why should we care about science books? *Journal of Science Communication*, **6**(1). Available online at: **http://jcom.sissa.it/archive/06/01/Jcom0601%282007%29C03/**.

■ REFERENCES

Barringer, F. (1998). Cancer-drug news puts a focus on reporters and book deals. *New York Times*, 8 May, p. 11.

Berkow, R. (1987). The *Merck Manual*: 'firm and faithful help'. *Book Research Quarterly*, **3**, 56–9.

Blackett, P.M.S. (1965). Introduction. *Philosophical Transactions of the Royal Society of London*, 258A, vii–x.

Boytinck, P.W. (1980). *C.P. Snow: a Reference Guide*. G. K. Hall, Boston.

Brenner, S. (2000). The house that Jim built (review of J. D. Watson, *A Passion for DNA*). *Nature*, **405**, 511–12.

Brockman, J. (1995). *The Third Culture*. Simon and Schuster, New York.

Brockman, J. (2002). *The Next Fifty Years: Science in the First Half of the Twenty-first Century*. Vintage Books, New York.

Brockman, J. (2006). *What We Believe but Cannot Prove: Today's Leading Thinkers on Science in the Age of Certainty*. Harper Perennial, New York.

Brockman, J. (2007). *What is Your Dangerous Idea? Today's Leading Thinkers on the Unthinkable*. Harper Perennial, New York.

Comfort, N.C. (2001). *The Tangled Field: Barbara McClintock's Search for the Patterns of Genetic Control*. Harvard University Press, Cambridge, MA.

Crellin, J.K. (1979). Mrs. Marcet's *Conversations on Chemistry. Journal of Chemical Education*, **56**, 459–60.

Davidson, K. (1999). *Carl Sagan: a Life*. John Wiley, New York.

Dean, C. (2007). James Watson quits post after remarks on races. *New York Times*, 26 October, p. 18.

Dennett, D.C. (2006). *Breaking the Spell: Religion as a Natural Phenomenon*. Viking, New York.

Dolan, B. (2000). The language of experiment in chemical textbooks: some examples from early nineteenth-century Britain. In: *Communicating Chemistry: Textbooks and their Authors* (ed. A. Lundgren and B. Bensaude-Vincent), pp. 141–64. Science History Publications, Canton, MA.

Fayard, P. (1988). *La Communication Scientifique Publique: de la Vulgarisation a la Mediatisation*. Chronique Sociale, Lyon.

Finley, F. and Pocovi, M.C. (1999). *Science Textbooks: Textbook Contributions to Science Curriculum Reform*. Unpublished manuscript. University of Minnesota.

Fischer, C.S. (1996). *Inequality by Design: Cracking the Bell Curve Myth*. Princeton University Press, Princeton, NJ.

Fraser, S. (1995). *The Bell Curve Wars: Race, Intelligence, and the Future of America*. Basic Books, New York.

Harris, S. (2006). *Letter to a Christian Nation*. Knopf, New York.

Henderson, A. (1997). Grey literature and publishing research [editorial]. *Publishing Research Quarterly*, **13**, pp. 3–4.

Henig, R. M. (2002). The life and legacy of Paul de Kruif. *APF Reporter*. Available at: **http://www.aliciapatterson.org/APF2003/Henig/Henig.html**.

Henneken, E.A., Kurtz, M.J., Accomazzi, A., Grant, C.S., Thompson, D., Bohlen, E. and Murray, S.S. (2007). Staying up-to-date in the planetary sciences with myADS. *AAS/Division for Planetary Sciences Meeting Abstracts*. (Abstract available at: **http://adsabs.harvard.edu/abs/2007DPS....39.2708H**.)

Hess, H.H. (1962). History of ocean basins. In: *Petrologic Studies: a Volume in Honor of A. F. Buddington* (ed. A.E.J. Engel, H.L. James and B.L. Leonard), pp. 599–620. Geological Society of America, Boulder, CO.

Holbrow, C. (1999). Archaeology of a bookstack: some major introductory physics texts of the last 150 years. *Physics Today*, March, 50–6.

Horrigan, J. (2006). *The Internet as a Resource for News and Information About Science*. Pew Internet and American Life Project, Washington, DC.

Hunt-Grubb, C. (2007). The elementary DNA of Dr Watson. *The Sunday Times Magazine*, 14 October, p. 24.

Kaufmann, W. (1995). Annual Reviews Inc., a saga of success. *Publishing Research Quarterly*, **11**, 80–9.

Kincheloe, J.L., Steinberg, S.R. and Gresson, A.D. (1996). *Measured Lies: the Bell Curve Examined*. St Martin's Press, New York.

Knight, D. (2000). Communicating chemistry: the frontier between popular books and textbooks in Britain during the first half of the nineteenth century. In: *Communicating Chemistry: Textbooks and Their Authors* (ed. A. Lundgren and B. Bensaude-Vincent), pp 187–205. Science History Publications, Canton, MA.

Leavis, F.R. (1963). *Two Cultures? The Significance of C.P. Snow*. Pantheon Books, New York.

Lemann, N. (1997). The bell curve flattened. *Slate Magazine*, 18 January. Available at: **http://www.slate.com/id/2416**.

Lessing, L. (1963). The three ages of science writing. *Chemical and Engineering News*, 6 May, pp. 88–92.

Lewenstein, B.V. (1987). Was there really a popular science 'boom'? *Science, Technology and Human Values*, **12**, 29–41.

Lewenstein, B.V. (1995). From fax to facts: communication in the cold fusion saga. *Social Studies of Science*, **25**, 403–36.

Lewin, R. (1992). *Complexity: Life at the Edge of Chaos*. Macmillan, New York.

Lindee, M.S. (1991). The American career of Jane Marcet's *Conversations on Chemistry*, 1806–1853. *Isis*, **82**, 8–23.

McElheny, V.K. (2003). *Watson and DNA: Making a Scientific Revolution*. Perseus Publishing, New York.

Madden, M. and Fox, S. (2006). *Finding Answers Online in Sickness and in Health*. Pew Internet and American Life Project, Washington, DC.

Maddox, J. (1988). The big Big Bang book. *Nature*, **336**, 267.

Mellor, F. (2003). Between 'fact' and 'fiction': demarcating science from non-science in popular physics books. *Social Studies of Science*, **33**, 509–38.

Morrison, R.T. and Boyd, R.N. (1959). *Organic Chemistry*. Allyn and Bacon, Boston, MA.

Myers, G. (1990). *Writing Biology: Texts in the Social Construction of Scientific Knowledge*. University of Wisconsin Press, Madison, WI.

Ortolano, G. (2002). Two cultures, one university: the institutional origins of the 'two cultures' controversy. *Albion: a Quarterly Journal Concerned with British Studies*, **34**, 606–24.

Pauling, L. (1947). *General Chemistry; an Introduction to Descriptive Chemistry and Modern Chemical Theory*. W. H. Freeman, San Francisco.

Poundstone, W. (1999). *Carl Sagan: a Life in the Cosmos*. Henry Holt, New York.

Rodgers, M. (1992). The Hawking phenomenon. *Public Understanding of Science*, **1**, 231–4.

Romano, C. (1995). Agent of the eggheads. *Philadelphia Inquirer*, 5 September, p. D01.

Samuelson, P.A. (1948). *Economics, an Introductory Analysis*. McGraw-Hill Book Co., New York.

Sears, F.W. and Zemansky, M.W. (1947). *College Physics*. Addison-Wesley, Cambridge, MA.

Segerstråle, U.C.O. (2000). *Defenders of the Truth: the Battle for Science in the Sociobiology Debate and Beyond*. Oxford University Press, Oxford.

Shinn, T. and Whitley, R. (eds) (1985). *Expository Science: Forms and Functions of Popularization*. Reidel, Dordrecht.

Snow, C.P. (1959). *The Two Cultures and the Scientific Revolution*. Cambridge University Press, Cambridge.

Snow, C.P. (1963). *The Two Cultures, and a Second Look*. Cambridge University Press, Cambridge.

Stenger, V.J. (2007). *God: the Failed Hypothesis. How Science Shows That God Does Not Exist*. Prometheus Books, Amherst, NY.

Topham, J. (1998). A textbook revolution: J. Deighton and Sons and the reform of mathematics in early nineteenth-century Cambridge. *Joint Meeting Between the Textbook Colloquium and the British Society for the History of Science*. Available at: **http://www.open.ac.uk/Arts/TEXTCOLL/ papers_toc.html**.

Turney, J. (2006). And the winner is . . . *Science Books Blog*, 20 October. Available at: **http://sciencebooksblog.blogspot.com/search?q=and+the+winner+is**.

Waldrop, M.M. (1992). *Complexity: the Emerging Science at the Edge of Order and Chaos*. Simon and Schuster, New York.

Watson, J.D. (1965). *Molecular Biology of the Gene*. W. A. Benjamin, New York.

Werskey, G. (1978). *The Visible College*. Holt, Rinehart and Winston, New York.

Werskey, P.G. (1971). British scientists and 'outsider' politics, 1931–1945. *Science Studies*, 1, 67–83.

Wilson, E.O. (1975). *Sociobiology: the New Synthesis*. Belknap Press of Harvard University Press, Cambridge, MA.

Yoxen, E. (1985). Speaking out about competition: an essay on the double helix as popularisation. In: *Expository Science* (ed. T. Shinn and R. Whitley), pp. 163–81. Reidel, Dordrecht.

■ FURTHER READING

- Gregory, J. (2005). *Fred Hoyle's Universe*. Oxford University Press, Oxford. Several studies have been done on the interactions between physics and books. Jane Gregory's biography of Fred Hoyle is a case in point. In this book she pays special attention to his use of books to argue for scientific ideas that were not gaining support in the traditional scientific literature.

- Leane, E. (2006). *Reading Popular Science: Disciplinary Skirmishes and Textual Strategies*. Ashgate, Aldershot. This book is a recent example of a literature-studies approach to understanding the structure of popular science books.

- Harmon, J.E. and Gross, A.G. (2007). *The Scientific Literature: a Guided Tour*. University of Chicago Press, Chicago. Many scholars have looked at the rhetoric of science paying special attention to the role books have played in science across historical times. This book is a recent example.

■ USEFUL WEB SITES

A number of sites provide collections of reviews of science books, including:

- Scibooks.org: http://www.scibooks.org/
- The Naked Scientists: http://www.thenakedscientists.com/HTML/books/
- Children's Science Book Review: http://www.librarians.info/CSBR/index.html
- *Science*: http://www.sciencemag.org/books/

5.2

Science communication in fiction

Jon Turney

Reading science

Story-telling is fundamental to communication. Indeed, some working at the interface between literary criticism and cognitive science have argued that stories are the building blocks of human thought, our best aid to understanding the world (Turner 1996). So it seems natural to ask whether the kind of story-telling we find in fiction can contribute to science communication. The answer, I think, is that it can. But not necessarily in any straightforward way. The contribution is not straightforward because good fiction texts are very complex creations, and can and will be read in many different ways. And they are highly diverse, so will lend themselves to many different uses. But many good fiction writers are expressing the fruits of a curiosity about the world which extends to science and technology as important features of contemporary culture. Their work certainly provides resources to consider in any comprehensive approach to science communication. What kind of resources? Well, the universe of fiction is large and this chapter can only cover a small sample. It is also limited by my own personal reading, and by the fact that there are very few studies of contemporary fiction which features science or scientists— as opposed to science fiction. Such attention as literary critics have given to recent science has tended to lead to more general arguments than those concerned with specific depictions of science—such as Katherine Hayles' provocative reading of Doris Lessing's *The Golden Notebook* in the light of chaos theory (see Hayles 1989). As Lessing's novel pre-dates the general awareness of chaos theory, this is a piece of involved textual analysis rather than a direct commentary on science in fiction.

This lack of scholarly analysis may change, as the bulk of literary work which includes portrayal of science or scientific ideas is relatively recent. Although there are enough examples from the 19th century to offer some insights into the growth of the scientific profession then (see Russell 2007), there were relatively few appearances of scientists as fictional characters in the first three-quarters of the 20th century, again excepting science fiction. The explosion of science since World War II (post-1945), and the somewhat later expansion of popular science as non-fiction, have encouraged authors to develop their interest in science and scientific affairs in recent decades.

In this light, in what follows I offer some general considerations about fiction and science, then look at some instructive examples in more detail. I conclude by asking whether the goals of novelists in relation to science—or how their books might be read in terms of science communication—have changed as some aspects of recent science have come to permeate wider culture more generally.

What use is a novel?

Science communication is generally pursued in the hope that someone will learn something—whether actual facts and concepts, ideas about how science works, or about science, technology and society.[1] Few people, it is safe to assume, pick up a novel hoping to benefit from this kind of learning. We read fiction, I suppose, to be entertained, beguiled, moved or provoked. If it is any good, maybe we can learn from it too. What do we learn? Something about the range of human types, perhaps, about how people behave, how lives are or were lived. In the grandest formulation, literature conveys insights into the human condition.

Of course the people depicted in a novel or a play can be doing anything at all—the flexibility of the forms is part of their appeal. And if they are involved in science, readers may also learn things about that in the course of the story. So in principle, it is possible for fiction to play its part in communicating science. There is, of course, an entire genre which is relevant here, but the possibilities go well beyond science fiction (which may or may not relate to real science at all). We can find science and scientists in historical novels like Clare George's *Cloud Chamber*, fictionalized biographies like David Leavitt's *The Indian Clerk* based on the life of the early 20th century Indian mathematician Srinivasa Ramanujan, and in an increasing number of contemporary novels which engage with science such as William Boyd's *Brazzaville Beach* with its combination of primatology and mathematics or Jeanette Winterson's *Gut Symmetries*, a rather puzzling exploration of physicists' lives.

As I've suggested, fiction is so various that generalization is always likely to be foolhardy. But with that caveat, there are things fiction can communicate at least as successfully, in some cases, as 'non-fiction'. The distinction, thus stated, is easily understood, but not always clear-cut in practice. As the term implies, non-fiction is defined by what it is not —it is not made up. But while most texts are relatively easy to place on one side of the fiction/non-fiction divide, there are plenty more which blur the boundary. Some do so by incorporating elements of fiction and non-fiction in the same work. This may be simply a matter of providing dialogue, or even private thoughts, in a biography or a popular history when it is often clear that the author could never authenticate the actual words, let alone the putative thoughts. More problematic is when fiction is passed off as non-fiction, sometimes playfully, sometimes with more questionable intent. I am not going to discuss this any further in general, but comment further in some of the examples below.

1. When the communication project embraces interactive or dialogic forms, of course, the learning can be more than one-way, but although novels are immersive they are not fully interactive in this sense.

Facts in fiction

A novel is not a textbook, so one might imagine that fiction is not generally good for conveying factual or conceptual/theoretical content of science. It may of course incorporate such material as it were incidentally. But any writer who sets out to convey a large slab of science in a work of fiction faces several problems. First, does it matter to the reader whether this science is accurate? As the critic and novelist Adam Roberts remarks, 'Application of conventional scientific orthodoxy as a criterion of judgment for an aesthetic object is fundamentally foolish even when applied with absolute consistency' (Roberts 2005, p 16). Roberts is writing here about science fiction, but his suggestion applies equally well to non-genre fiction. And this is only one strike against science in fiction. Incorporating large amounts of science in a work of fiction is liable to violate the contract with the reader, who is seeking a story, not instruction. And it creates technical problems for the writer—the need for exposition means that characters tend to end up lecturing one another about things they ought to know already. This interruption of the narrative for an infodump can happen for other reasons too, but it may be more noticeable if the information being dumped is about science. In addition, as I discuss below, it can create a temptation to pass off fiction as fact when an author has an axe to grind.

Still, there are examples of basic science being presented as exposition successfully. The mini-lecture about DNA given to the park visitors in *Jurassic Park* by the medically trained Michael Crichton (rendered in the film of the book as a rather cute animation) is a fairly standard exposition of a basic piece of science. Less successful, though, is the same book's treatment of chaos theory, although it has been analysed sympathetically as an exposition of this new area of inquiry. Change the context and a similar piece of exposition can sit oddly in a fictional text. Carl Djerassi, celebrated as the developer of the contraceptive pill, has lately turned to fiction and *The Bourbaki Gambit* (1996) is one of a series of novels and plays he has written to display how scientific work gets done and illuminate dilemmas of scientific application—usually in biology. In this novel the plot is about ageism in science and turns on a group of biologists who are regarded as past it who band together and publish new work anonymously (as did the celebrated school of French mathematicians who took the pseudonym Bourbaki). All is well until one of them makes a major discovery, which turns out to be the polymerase chain reaction (PCR) used for amplifying small samples of DNA.

But PCR is a real-life technology, with a real-life—and quite famous—inventor in Kary Mullis. He has even written his own autobiography. So it seems rather confusing having the same technique, which is described and explained at some length, used in the novel in an otherwise fictional context.

This confusion seems harmless, except perhaps to the novel concerned. More problematic is the confusion deliberately created in two of Michael Crichton's more recent novels, *State of Fear* and *Next*. The first focuses on climate change, and Crichton's (real world) conviction that the science has been misrepresented by eco-fanatics. The novel is a clumsy effort, a thriller with characters who interrupt the action to give long lectures about the 'true' state of climatic affairs. Like its successor, *Next*, about biotechnology, it also comes with notes and bibliography and a brief statement of Crichton's own conclusions about the issues it highlights—both oddities in a work of fiction. In addition, in the main text he incorporates a wide range of news reports, articles and memos from relevant organizations. Some of these are genuine, some are concocted by the author—but he declines to make clear which are which.[2]

This suggests, at the least, that the novel itself is in this case a poor vehicle for science communication. This does not mean that it cannot be used indirectly. Some of the scientists Crichton took advice from, for example, published a detailed analysis of his errors at the RealClimate web site (http://realclimate.org/). So this particular piece of fiction has prompted debate in ways which go beyond regular reader responses—and which bears comparison with the discussion of other climate fictions like the blockbuster Hollywood catastrophe movie *The Day After Tomorrow* or Bjorn Lomborg's non-fiction opus *The Skeptical Environmentalist*.

These two novels mark an unusual extreme in blurring the boundary between fiction and non-fiction and, I would argue, one likely to be self-negating. Readers are likely to sense that the author is not being entirely open about what is fact and what is made up, and lose confidence in the whole enterprise. It is also relevant to note that, in my view, these are bad novels, in which such niceties as characterization are overwhelmed by the author's sense of mission. However, Crichton's procedure isn't new, nor is it necessarily disastrous. Margaret Atwood's near-future catastrophe novel *Oryx and Crake* makes much of contemporary biotech news, and you can still look up the headlines which she wove into her story on the book's US web site—though there wasn't actually one about a scientist who decides to destroy humanity using a genetically engineered virus because, well, he can, as featured in the book. Further back, the well-regarded British science fiction writer John Brunner showed Atwood how it could be done more than 30 years earlier in *Stand on Zanzibar*. His arresting picture of a possible future, inspired in part by John Dos Passos' 1930s vision of the USA, mixes real and made-up news reports, letters, diaries and conventional narrative. Both Atwood and Brunner's books are highly readable, but neither has a specific agenda in the manner of Crichton—they are more concerned to portray possible outcomes of a whole complex of current trends, scientific and social.

Another respect in which Crichton, Brunner and Atwood are all similar is that their works are, first and foremost, books. Although the more recent novels may be discussed on the web, their own web sites, when they exist, are chiefly for conventional publishers' promotion, not discussion of the issues they may raise. The newer interactive media afford plenty of scope for blurring the boundaries of the novel and involving readers more closely, but these possibilities remain largely undeveloped for now.

Returning to the traditional novel, then, other notable uses of fiction related to bio-technology include David Rorvik's largely forgotten book *In His Image*, which comes into this chapter because it was really a novel but presented and published as non-fiction in the late 1970s. Rorvik, a successful science reporter, portrayed the first human clone being born at the behest of a wealthy egomaniac seeking immortality. And biologist Lee Silver used fictional vignettes of people grappling with future genetic technologies

2. In which light, his insistence on the need for a clean separation between science and politics reads a little oddly—see his speech on 'Environmentalism as Religion'; http://www.michaelcrichton.com/speech-environmentalismaseligion.html.

all through his non-fiction *Remaking Eden* in 1998. Like Rorvik, Silver is a poor fiction writer (though an interesting writer), so his little fictions, though clearly framed as stories to think with, fell a bit flat.

A summary conclusion from these examples is that you need to be a really good novelist to bring off this kind of flirting with the edges of non-fiction. Atwood qualifies, and although *Oryx and Crake* is not her best work it is undeniably full of memorable images of a disastrous future. Crichton is, at best, an efficient thriller writer, but his current work is hobbled by his compulsion to preach to the reader. Even by his own standards, those of a popular thriller writer, I would argue that his recent work is not up to his best. His work certainly suffers in comparison with more subtle treatments of biotechnological possibilities, such as Kazuo Ishiguro's allegorical *Never Let Me Go*, or recent literary novels in which biotech is just a taken for granted feature of the landscape, like Zadie Smith's *White Teeth* or David Mitchell's *Cloud Atlas*. But none of these is remotely concerned with conveying facts about science, which rather underlines the generalization which began this section.

Science in context

Facts and concepts aside, what fiction may at times do better at than non-fiction is just what we might expect. Novelists need to inhabit the minds of their characters, and can help us do the same. This requires an act of imagination, and can go further than is usually possible in non-fiction. Fiction can thus communicate features of scientific mentalities and milieus, as well as aspects of context and relations between science and society which are more difficult to convey effectively outside of a created story. The flexibility in point of view afforded by fiction, and the potential to write as an omniscient narrator who can see inside people's heads, are not completely barred to the non-fiction writer but are harder to realize convincingly.

One of the most developed examples of how this can work well is a book which shares many features of non-fiction, including extensive scholarly research, notes and a bibliography. But historian Russell McCormmach's *Night Thoughts of a Classical Physicist* (1982) is a work of fiction, depicting a thoroughly convincing German physicist contemplating his not very distinguished career in the closing months of World War I (1914–18). McCormmach's Viktor Jakob, a poignant figure meditating on the end of an era in science and politics, is a composite of historical figures, and many real physicists appear in the book. But the author's achievement is literary as well as historical. It is a short, intense book which stays in the mind, as a good novel should.

Compared with this, a dramatic treatment of a historical episode in non-fiction—such as Dava Sobel's best-selling *Longitude* about the clock-maker John Harrison—is liable to be a less rich depiction of events than a novelist's version of a particular historical episode.[3]

---

3. Though, as so often in this kind of discussion, spelling out the generalization causes exceptions to spring to mind right away. In this case, Carl Zimmer's *Soul Made Flesh* is an excellent, non-fiction re-creation of the Oxford milieu which helped nurture the embryonic Royal Society, as well as witnessing the first scientific investigation of brain anatomy.

Such works can be closely based on known facts, but with added imaginative elements. Clare Dudman's novel *Wegener's Jigsaw* is an interesting example, as she began work intending to write a biography of Alfred Wegener, the first proponent of continental drift. But she found herself impelled to re-create episodes in his life in ways which went beyond the possibilities of conventional biography, and produced a novel which is a closely researched account of Wegener's life and work. The voice which narrates the book is based on the writing in Wegener's diaries, but relates scenes and thoughts which he did not record, including his death on an Arctic expedition. The result is, perhaps, a hybrid work, but publishers, bookstores and readers need categories so it is published as a novel, and a very good one. The reader learns from the back matter that Wegener was a real historical figure, and that the author researched his life extensively, but while this is interesting to know, it does not much matter it seems to me, when the book is being read for its story.

Similar in some ways, although treating much more familiar material, is the late Harry Thompson's long novel *This Thing of Darkness*, exploring the relation between Charles Darwin and Robert Fitzroy, captain of the *Beagle* and the austere commander who took the young Darwin as his gentleman companion on their round the world voyage. Again, the novel has drama which cannot be so effectively treated in non-fiction, as Fitzroy's post-*Beagle* career (trying to convince his masters in the Admiralty of the value of meteorology, his recurrent depression and eventual suicide), is well-furnished with known facts but provides little clue to his troubled inner life.

Less successful than these, in my view, is cosmologist Janna Levin's *A Madman Dreams of Turing Machines*. Levin followed her unconventional popular science book *How the Universe got its Spots*, a diary presented as unsent letters to her mother, with a dramatization of episodes from the lives of logicians Kurt Gödel and Alan Turing. So the reader witnesses Alan Turing trapped beneath the floorboards of Sherborne School by his bullying classmates; Gödel bringing the devastating news of 'undecidability' to the smoke-wreathed café table of the Vienna Circle of philosophers in the 1930s; Turing at Bletchley Park during World War II, wrestling with the Enigma codes and spurred to mechanize computation; Gödel in Princeton, declining into paranoia and self-starvation; and Turing in Manchester, harassed for his open homosexuality and eating his last, cyanide-laced, apple.

All of these make compelling reading, as anyone who has read biographies of these two intriguing figures can testify. Their lives have an enduring fascination, and Turing's has already been turned into a successful play. But the result here is a slightly unsatisfactory book. The two life stories are told, interwoven and largely unadorned, in conventional chronological order. Yet the point of bringing them together is not quite clear. And there is little drama, in the sense of characters who are changed and whose changes we now understand. They just proceed with their lives to their seemingly preordained conclusions.

At the same time, the effort of re-creation does not leave much space for exposition, so the reader appreciates the real achievements of the two subjects only tangentially. Gödel's work comes across a little better, perhaps because it was less diverse than Turing's mixture of philosophy, logic, mathematics, engineering design and, at the end of his life though not featuring in the book, biology.

One might argue that this book is an unsteady hybrid because it doesn't take the possibilities of mixing genres quite far enough. Levin's prose achieves something of the quality of physicist and writer Alan Lightman's poetic meditations in his book of short stories *Einstein's Dreams*, but that is much more clearly in a realm of metaphysical fiction. On the other hand, the mathematician and prolific popular science writer John Casti's *Cambridge Quintet*—also described by the publishers as a novel—explores similar territory to Levin's book by bringing together C. P. Snow, Erwin Schrödinger, J. B. S. Haldane, Turing and Wittgenstein at a fictional dinner party. Casti's fiction is threadbare—he simply stages a dinner party conversation—but the exposition, much of it drawn from his characters' own writing, went deep into the ideas.

Other more successful historical re-creations which place science, and scientific figures, at the heart of a fictionalized narrative include the distinguished Irish novelist John Banville's two fine early historical novels *Kepler* and *Copernicus*, and Roger Macdonald's *Mister Darwin's Shooter*. The latter bears close comparison with *This Thing of Darkness*, and tells the story of Simon 'Syms' Covington, Darwin's shipboard assistant on the *Beagle* and house servant. Through the relation between the two men, which continued through correspondence after Covington settled in Australia, Macdonald dramatizes the crisis of faith in Victorian culture. Darwin loses his faith, and Covington agonizes about the atheistic book, *On The Origin of Species*, which he has helped make possible by his energetic collection and preservation of specimens for his former master. The unusually close master–servant relationship, which Darwin hints at in various letters, is an excellent vehicle for reimagining this conflict of ideas, but it was more of a feat of imagination in this case, perhaps, than in Thompson's novel about Fitzroy. Very little is actually known about Covington's life or character—he was a servant after all—and as the author says, 'I interpreted Covington for fictional purposes by taking the known facts of his life into the realm of speculation' (MacDonald 1998).

LabLit and its limitations

There are a fair number of novels with a more recent setting than *Mr Darwin's Shooter* which depict the ways of scientists. I have already mentioned Carl Djerassi's novels and plays, which explicitly portray the doings of the scientific tribe, and are indeed described by one critic as 'a kind of fictional cultural anthropology' (Grünzweig 2004).[4]

But there are quite a few other novels which feature laboratory or institutional settings, a type dubbed 'LabLit' by its fans, and which has spawned a web site devoted to promoting such works as a way of making more or less plausible pictures of scientific work accessible to wider publics. This genre, if such it is, is one which 'depicts realistic scientists as central characters and portrays fairly realistic scientific practice or concepts, typically taking place in a realistic—as opposed to speculative or future—world', according to the website leading this discussion.[5]

The LabLit site offers other interesting material, some of it non-fictional, but is clearly in part an extended statement that fiction can be a useful aid to science communication. It is a little difficult to know what to make of this though. The idea that realistic depiction

4. For an extended critical discussion of Djerassi's literary efforts see Grünzweig (2004).

5. http://www.lablit.com.

of scientists is a quality to seek out for its own sake, as it were, seems pretty limited—and of course the books proposed need to have other winning qualities as well to be worth reading. An old example of 'LabLit', for example, William Cooper's *The Struggles of Albert Woods*, is a wry and witty reflection on English social and intellectual life in the 1930s, from the viewpoint of the 1950s, which happens to have a protagonist who is a not very gifted chemist on the make. It is still worth reading today for its literary quality, not because it happens to feature scientific life at Oxford.

More recently, and on the 'realistic' science fictional side of the genre, Kim Stanley Robinson's *Forty Signs of Rain* could be read for a detailed depiction of life inside the bureaucracy of a near future US National Science Foundation (NSF)—one of the first science policy fictions, perhaps. But this is surely incidental to the author's main intent, and most readers' main interest, which is to explore the consequences of global warming, and how we might respond to it, politically, socially and technologically.

This is just one example of the limitations of inspecting novels and stories through the filter of a search for 'LabLit'. There are always going to be more worthwhile books which are 'about' something else—perhaps many other things—but happen to feature science in some important way, than there are rewarding reads in which science is the main business depicted. It would be a stretch, for example, to classify Neil Belton's fine, if slow-moving, novel *A Game with Sharpened Knives* as LabLit. It is in some ways reminiscent of McCormmach's *Night Thoughts of a Classical Physicist*, albeit its protagonist is the elderly refugee Erwin Schrödinger trying to do theoretical physics in wartime Dublin. But while one does absorb some discussion of 20th century physics, and the problems of physical understanding, Belton is just as interested in the condition of the newly fledged Irish Free State as he is in quantum mechanics.

Similarly, the novels by Zadie Smith, David Mitchell or Kazuo Ishiguro which I have already mentioned, either incorporate science as one element in a complex narrative which is being stretched to embrace a whole culture or use it to underpin an allegorical story—in Ishiguro's case about human difference. But even then we have only started to sample the richness of the novel as a vehicle for reflecting on science. For between the determinedly pedagogic novels of a Crichton or a Djerassi and the literary novels in which science is treated in passing as a part of the culture there are a range of works in which scientific ideas are represented, figured or discussed in various ways which help animate the narrative.

There are well-known theatrical examples which can be mentioned here, such as Michael Frayn's *Copenhagen* or Tom Stoppard's *Arcadia*. But let me finish with a look at two novelists who are fascinated by the sciences, one from each side of the Atlantic, but who have used it in different ways in their literary careers.

Ian McEwan and Richard Powers—a different kind of knowledge?

Ian McEwan has often written of his sympathy for science, and also makes it plain that, in so far as a novelist is a student of human nature, study of scientists' ideas on the subject repays attention. He has a particular interest in neo-Darwinian approaches such as evolutionary psychology. Novels of his which pursue these possibilities explicitly are the exception. Nevertheless, the two which stand out from the rest in this respect are fascinating examples of how a novelist can blend scientific elements into a complex story. In *Enduring Love* he begins with a set piece in which a group of people hanging from the ropes of an ascending hot air balloon have to each decide whether to cling on as the balloon rises higher and higher—which maximizes the chances they might all survive— or whether to let go, saving themselves but increasing the hazard for those still clasping the rope. It is a gripping dramatization of the kind of situation described by game theory, in which the payoffs one player receives are affected by others' actions, and game theory has been one important way of formalizing certain aspects of evolutionary theory—a fact underscored by the book's narrator, who happens to be a popular science writer.

Also central to the plot is a character who believes he is in love with Joe, the narrator— the result, he decides, of a neurological syndrome which is described in a scientific paper appended to the main text. A number of readers and reviewers took this to be genuine, and there was general amusement at their expense when the author revealed that, like the whole book, it was made up.

Joe's fascination with our understanding of the mind, and the limits of what science can explain about behaviour is shared by the lead character in the same author's later novel, *Saturday*. Set on the day of the exceptionally large but ineffectual London demonstration in February 2003 against the imminent Western invasion of Iraq, the novel explores the world of a neurosurgeon, Henry Perowne, and his family. Their comfortable life is thrown into relief, not just by the turbulence in the wider world but by an encounter with a thuggish criminal. But our neurosurgeon quickly perceives that he is showing early signs of Huntington's disease. Later, when the same man is injured after invading Perowne's dinner party, the surgeon retires to bed, but is aroused by a summons to the hospital—the man needs emergency brain surgery to save his life. Perowne obliges, fulfilling his moral obligation as a doctor, but also knowing that the life he saves is destined to end horribly from the incurable genetic degeneration decreed by Huntington's.

If this capsule summary makes the plot sound a tad implausible, well, it is. But the skill with which Perowne's world is rendered and the way the author makes the events of the day unfold disguises this quite effectively until near the end of the book. But the point here is that, again, one of our leading novelists incorporates a range of scientific ideas and materials—he spent much time mugging up in neurology—in pursuit of a novelistic meditation on what shapes a life, and whether we deserve our fates, good or bad.

My second candidate for a 'scientific novelist' is the enormously talented American Richard Powers. Powers studied initially as a physicist, before switching to English literature, and worked as a computer programmer, so it is no surprise to find scientific elements in many of his books. They range from a deep exploration of molecular genetics and coding in the monumental *The Goldbug Variations* to the presence of a rather Einsteinian Jewish physicist as the father of a mixed-race family whose story is told in the magnificent panoramic exploration of race, art and America in *The Time of Our Singing*.

But the most interesting perhaps, are the pair of novels which feature a direct engage-ment with neuroscience. In *Galatea 2.2*, published in 1995, the protagonist is a novelist, called Richard Powers, who encounters a university neuroscientist who insists he can program an artificial intelligence to pass an advanced oral exam in English literature. This elaboration of the Turing test, designed to establish whether a computer program might convincingly emulate human intelligence, prompts a great deal of discussion of computers, brains and the nature of humanity. The novel has some 'LabLit' elements, to be sure, to the extent that a reviewer in a cognitive science journal declared it to be a thinly disguised portrayal of the real-life Beckman Institute and that the characters were instantly recognizable to others in the field. This is, presumably some testament to Powers' skill at invention, as he assures us that all the characters, and their institute, were completely made up.

But he has a deeper intent than simply depicting scientists, once again on display in his latest novel, *The Echo Maker*. This features a young man who has suffered brain damage after an accident and exhibits a rare syndrome in which he recognizes everyone but his closest relative, his sister, who he takes to be an imposter. She enlists the help of a neuroscientist and popular author who studies such conditions, a character who has a certain amount in common with both Gerald Edelman and Oliver Sacks.

The book is a mystery story but, as with McEwan, part of the mystery is how much we can know about ourselves, and to what extent the brain makes up what appears to us to be the incontrovertibly real world. The point, for Powers, is that the novel as an aid to introspection should embrace the new sciences of introspection. At the same time, he maintains, the sciences need fiction to bring out their full significance. As he told the *Paris Review*'s interviewer:

Once you read the neurological research, you know these guys are bursting with excitement about the implications of their work. But they can't get into all the implications because the implications don't come out of well-formed questions and they're not all answerable by reductive, empirical programs. Somebody has to come along and spin a story around it so we can take it up into our lives—so that we can figure out what it means in the grossest possible sense. There are places that empiricism simply can't get to.

This is where literature becomes a kind of knowledge that is not only unique but is valuable.

Powers (2003, p. 26)

This is an eloquent summary of the real possibilities of the novel as an aid to exploring the meaning of science, and not just neuroscience. Others have similar aims in respect of genetics, quantum mechanics or cosmology—the most popular disciplines for literary explanation tend to map fairly closely onto those which feature most prominently in popular non-fiction about science. But, as with 'popular' science as a category, the ensemble of titles remains extremely diverse, and many are read by relatively few people. The rule that bestsellers are rare exceptions and that there is a long tail of works which reach small audiences by modern mass media standards applies to novels as surely as to any other sector of contemporary publishing.

This creates a difficulty in offering any strong conclusion about trends in this kind of writing, or about the effects it may have. The fact that there are many titles each reaching a few thousand or tens of thousands of readers suggests that any important effects draw on many different books. At the same time, the wide differences between books—the glory of the novel remains its flexibility and diversity of form and subject matter—make it much harder to say what these effects might be.

But I would take from Powers' framing of his ambitions for the novel the conclusion that at its best (and bearing in mind that quality is not necessarily related to sales) the contemporary novel of science is, as literature ought to be, a long way from any simple notion of science communication. On the other hand it can also serve to enlarge our notion of the possibilities of science communication, in the most expansive definition of the term. It would also be nice to think that, in some fashion, literature and science communication share their basic values. Again, Powers has a formulation which might inspire us to focus our efforts more effectively in either sphere: I've always believed a book functions best when it leaves a person more capable of living in the world' (Powers 2003, p. 23).

■ REFERENCES

Brier, S. (2006). Ficta—remixing generalised symbolic media in the new scientific novel. *Public Understanding of Science*, **15**, 153–74.

Djerassi, C. (1996). *The Bourbaki Gambit*. Penguin, New York.

Grünzweig, W. (2004). Science-in-fiction: science as tribal culture in the novels of Carl Djerassi. In: *Science, Technology, and the Humanities in Recent American Fiction* (ed. P. Freese and C.B. Harris), pp. 231–48. Verlag Die Blaue Eule, Essen.

Hayles, K.N. (1989). Chaos as orderly disorder: shifting ground in contemporary literature and science. *New Literary History*, **20**, 305–22.

McCormmach, R. (1982). *Night Thoughts of a Classical Physicist*. Harvard University Press, Cambridge, MA.

Macdonald, R. (1998). Evolution of a novel: *Mr Darwin's Shooter*. *Australian Humanities Review*, December. Available at: **http://www.lib.latrobe.edu.au/AHR/archive/Issue-December-1998/mcdonald.html**.

Powers, R. (2003). The art of fiction. *The Paris Review Interviews*, no. 175. Available at: **http://www.theparisreview.com/media/298_POWERS.pdf**.

Roberts, A. (2005). *The History of Science Fiction*. Palgrave, London.

Russell, N. (2007). Science and scientists in Victorian and Edwardian literary novels: insights into the emergence of a new profession. *Public Understanding of Science*, **16**, 205–22.

Turner, M. (1996). *The Literary Mind—the Origins of Thought and Language*. Oxford University Press, New York.

■ FURTHER READING

- Brier, S. (2006). Ficta—remixing generalised symbolic media in the new scientific novel. *Public Understanding of Science*, **15**, 153–74. Brier argues that Michael Crichton's novels are examples of a new 'genre', fiction based on a real scientific problem. This is explored using *Jurassic Park* as a case study.

- Hayles, K.N. (1989). Chaos as orderly disorder: shifting ground in contemporary literature and science. *New Literary History*, **20**, 305–22. A highly sophisticated approach to reading scientific influences on contemporary fiction, informed by literary theory and science studies. Hayles has also written extensively on cybernetics, computer science and literature.

- Roberts, A. (2005). *The History of Science Fiction*. Palgrave, London. A very readable and up-to-date one-volume history of science fiction, by an author who combines a day-job as an English literature academic in the UK with work as a well-regarded science fiction novelist.

- Russell, N. (2007). Science and scientists in Victorian and Edwardian literary novels: insights into the emergence of a new profession. *Public Understanding of Science*, **16**, 205–22. In this paper, Russell argues that 19th and early 20th century fiction can be a fruitful study for students, illustrating the impact of the rise of science on wider culture.

■ USEFUL WEB SITES

- **LabLit: http://www.lablit.com**. The site offers an impressive list of fictional 'LabLit' works, more than can usefully be discussed in this chapter, and divides them, a little arbitrarily, into novels, cross-over novels (defined as 'science fiction with particularly realistic scientists'), as well as films, plays and (interestingly few) TV programmes.

- *The Paris Review*—**the interviews: http://www.theparisreview.com/literature.php**. This site offers a rich resource for writers' reflections on their work, covering an impressive range of 20th century authors. Some content is subscriber only, but many interviews are available to all.

5.3

Speaking to the world: radio and other audio

Martin Redfern[1]

In order to understand the landscape of science radio broadcasting in the 21st century it is interesting to take a historical view. I am fortunate to have worked for the biggest and probably the best science audio broadcaster in the world for my entire working life, across a quarter of a century of spectacular changes. By analysing those changes, I hope to reach an understanding of where we are today. In so doing, I will draw on my experiences in producing science radio programmes for the BBC. I will focus particularly on the production of magazine programmes, features and news stories for BBC Radio 4 and the World Service, finishing with a look at current trends in news production and consumption of new forms of audio.

Before I start, a brief note on terminology. 'Radio', or wireless as it was initially known, is a term that refers both to the device through which broadcasts are received and to the content of those broadcasts. With the advent of digital technologies, the term radio is being challenged somewhat, in part by newer forms of audio, such as podcasting, that do not fall neatly under this term. It is possible to argue that the term 'multi-platform audio' more accurately reflects the current landscape for these various forms of audio. However, for the sake of clarity, and also because the vast majority of my professional life has been as a producer of audio for broadcast on radio, and the examples listed reflect this, I have used the term radio wherever possible in this chapter.

'Ancient' history

Specialist science audio broadcasting goes back half a century. In this respect, 1957 was possibly the most important year there has been for global science, and for science broadcasting to the world, pre-dating satellite television and the internet. It marked the

1. Any errors, opinions or cynicism expressed in this article can be attributed entirely to the author. Neither the Open University nor the BBC are in any way to blame.

beginning of the space age with the launch of the Sputnik satellite and the start of serious exploration of the inner space of the world's oceans and poles in International Geophysical Year. Technology was beginning to transform lives through transport, household appliances, national defence and of course public sector and commercial broadcasting.[2] In the UK, science programming was beginning on both the BBC's Home Service and its Overseas Services. Following its launch in September 1946, the BBC's Third Programme was running scripted talks by scientists and the European Service in English began broadcasting a weekly round-up unimaginatively titled *Science and Industry*. That programme, though its name changed to *Science in Action*[3] in 1967 continues on the BBC's World Service to this day and is probably the world's longest running radio science show, rivalling even Patrick Moore's *The Sky at Night* on television.

Over the next 25 years, the External Services of the BBC grew to include 42 languages, broadcasting around the world mostly on shortwave to an estimated weekly audience of around 150 million people. Reflecting what Harold Wilson famously described as the 'white heat of technology' of the 1960s, the science team grew from a single correspondent to, by the 1980s, a unit of 14 people: editor and assistant editor, three writers, three producers, five secretaries and a trainee—me. The structure and the nature of what we did then now seem pretty rigid and unchanging; little did we know just how much change lay over the horizon.

The three script writers simply wrote scripts. These were usually straight text and were made available duplicated on paper in English for any of the language services to use in translation as they wished. These scripts, along with all the stories that make up the news, were distributed around the building in cylindrical canisters blown through tubes by compressed air. I can remember in my previous job as a studio manager, working night shifts, being able to use a bed in the manager's office in the long gaps between programme commitments. You always knew if a big news story was breaking because you would be woken to the sound of stories wooshing through the pipes above your head.

The three producers in the unit produced programmes, the same regular programmes week in, week out. *Science in Action* was the flagship topical round-up of the week; *Discovery* was its in-depth sister programme, again half an hour long but often containing only two long interviews with relevant experts. Then there were *The Farming World* at 20 minutes and *Nature Notebook* at 10 minutes. The former was a wonderful programme making use of reporters who happened to be travelling, usually on someone else's budget, to see agricultural projects around the world. The producer realized that farmers in the developing world probably wouldn't be listening to an English-language overseas broadcaster and instead aimed the programme at the next level up of trainers, administrators and non-governmental organizations (NGOs). By sharing the experiences of practical projects around the world, we hoped that they picked up valuable tips and that the programme contributed profoundly to development. Finally, there was *Nature Notebook*, a programme that was conceived long before environmental journalism became mainstream and sometimes consisted of interviews conducted in whispers, often whilst watching some endangered creature.

2. For a discussion of the development of public sector radio broadcasting at the BBC see Eldridge *et al.* (1997).

3. The programme web site can be found at: http://www.bbc.co.uk/worldservice/programmes/ science_in_action.shtml.

There would not have been a trainee in the team had it been left to the accountants. The post of studio manager was seen as a general graduate recruitment programme for the rest of the BBC, so the science unit got free labour in exchange for a little training and my career prospects improved spectacularly.

In those days, transatlantic colleagues would often ask where I did my journalism masters degree. With a smile, I would reply 'the university of life'. There were very few media courses in the UK then and none that I was aware of in science communication. Rather, the BBC was seen as the training ground for broadcasters internationally. It was very much practical training in craft skills and front-line journalism. Things are very different today with full-time and part-time post-graduate science communication courses at several universities, not least the one linked to this publication. The result is a large pool of people looking for a job. From my experience, they are often well qualified academically, if lacking in practical experience.

The nature of radio

Radio is a very different medium from either print or television. It is highly personal to the listener(s), even intimate. Many people listen when they are on their own, at home or in the car and increasingly on personal headphones through a portable radio or MP3 player. Whilst music and chat stations are often simply used as background noise, speech networks on radio often command higher levels of attention than television. Although we hope there are millions listening, as a science radio producer I often like to think of there being one single listener, and work from there. It's almost as if I have been invited into his or her home, into a position of trust. That gives me important responsibilities not to waste the listener's time or abuse that trust and to explain what I have to say clearly, accurately and entertainingly; interviewees are invited to do the same. If I fail, either in my production or in my marshalling of interviewees, it is the listener not I who can reach for the 'off' button.

Radio is a linear medium. Even on a recording it's difficult to go back and listen to that last sentence again because you didn't quite understand it. On a live broadcast it's impossible. You have to get it right first time and be clear and understandable to everyone. That's not to say that you can't challenge the listener and expect an effort of understanding in return. The intimacy of the medium can bring a greater level of attention and make it quite possible to explain complex ideas. But it has to be clear and free of jargon. Personally, I think it's a big mistake to oversimplify and become too didactic. That is likely to alienate and insult listeners, many of whom have interests and knowledge about particular areas in science; so-called professional amateurs, or 'pro-ams' (see Leadbetter and Miller 2004 for discussion of this concept). Even if they have little or no knowledge of science, listeners don't like to be lectured. There are ways to include sufficient explanation without sounding didactic, principally by interviewers representing the listener and not trying to show off superior knowledge. My general rule is always to underestimate the knowledge of listeners but never to underestimate their intelligence.

The linear nature of audio brings some big advantages. It's great for storytelling. It can be far more effective to tell the story of how a discovery was made than simply to describe what that discovery is. Strangely, audio can also paint spectacular pictures.[4] Through the use of location recordings, sound effects and descriptive narrative, listeners not only see the picture in their minds but their attention is drawn to significant features and subtleties that they might miss on television. But in describing scientific research, there are pitfalls. There can be no tables and charts of statistics, no glossary, footnotes, references or index.

Good radio can be an art form. Through location atmosphere, relevant music and personal testimony, it's possible to create feelings and moods as well as pictures. A simple story told with feeling, preferably from personal experience, can be powerful and moving as well as informative. Some of the best radio programmes—and some of the hardest to make—invite listeners to participate in a poetic sound journey.[5] Even fact-packed science documentaries can sometimes dispense with the presenter and put listeners in direct touch with first-person stories.[6]

Getting to know the listener

One of the first programmes I ever made was as a student volunteer for hospital radio. It was a hospital for those suffering from mental illnesses, and I'm quite sure in retrospect that the patients had particular hopes, needs and limitations. But nobody told me that so I went off and made a one-hour programme about new technology at the Farnborough Air Show. Of course, this may have been light relief for some of the patients, but with hindsight it does not look like the most obviously relevant topic. Professional broadcasters have to take rather more care, and the BBC puts in a tremendous effort of time and resources trying to find out who is listening to what and when and what they think of it.

Such research suggests that the World Service audience is vast. It is estimated that (in 2007) 183 million people listened to something on the World Service in one of 33 languages every week. That comes down to around 45 million listening in English, and to any one programme it is less still, but still a figure to turn most contemporary television executives and newspaper editors green with envy, particularly given that television audiences are fragmenting in the multi-channel era, and hard-copy newspaper circulations are in long-term decline (Holliman 2008).

The biggest audience for the BBC's World Service internationally in English is in Nigeria, with other West African nations such as Ghana not far behind. Rising prosperity and access to television has pushed India out of second place, which is now taken by

4. For an example of how this can be done 'listen again' at:
   http://www.bbc.co.uk/radio4/science/fiveholesintheground.shtml.

5. For an example, 'listen again' at: http://www.bbc.co.uk/radio4/science/nature_20071001.shtml.

6. For an example, 'listen again' at: http://www.bbc.co.uk/radio4/science/smalldog.shtml.

the USA—perhaps surprising given that we don't broadcast there directly at all. It says something about changing trends in the technology of listening that more than 5 million Americans hear the World Service rebroadcast by other stations, picking it up direct from satellite or, in increasing numbers, listening over the internet.

The Cold War days of dissidents gathering in a dusty attic to listen in secret to *London Calling* over a crackly short-wave radio set are long gone. In this high-tech age, the listener may have hundreds of stations to choose from on FM, digital audio and online: the emergence of multi-platform audio discussed at the start of this chapter. Deregulation and the end of state broadcasting monopolies mean that the BBC is no longer one of only a few broadcasters offering uncensored news (for a discussion of public sector and commercial radio broadcasting see Franklin 1997), so we have to try much harder to attract and retain listeners with the highest standards of news, information and pro-gramming. And we have to be there loud and clear on the tuning dial that also carries the local pop music stations. As a result, World Service can now be heard on FM in 151 capital cities of the world.

A few years ago, audience researchers identified three distinct groups of BBC World Service listeners:

- the 'information poor' who are desperate for any news and who have very little education in science but are keen to learn;
- 'aspirational listeners', mostly in cities, who are keen to improve their general knowledge and enjoy being able to talk about everything from politics to quantum physics; and
- a more mature, well-educated audience who find television sometimes trivial and who want to engage with programmes on a deeper level.

In the context of the BBC's World Service, the first group are hardest to reach and may not understand English to a conversational level. Whilst it is all they can get, the aspirational audience love the BBC, but previous experience also suggests that they are likely to desert us as soon as they get television. The third group is older and smaller but more loyal and now growing more rapidly. Somehow we have to make science pro-grammes that will appeal to all three groups.

In the UK, the BBC's Radio 4 has more than 9 million regular listeners. Around 2.5 million hear at least one of our regular weekly science programmes. They tend to be older than the average radio listener, though there is an increasing young audience as well. Regular listeners sometimes time their day by the programme schedule, rising to the *Today* pro-gramme of news and current affairs and perhaps winding down with *A Book at Bedtime*. It is therefore not that surprising that they are very resistant to change.

Computerized scheduling and switching of programmes across eight different time zones enables many BBC World Service listeners across the world to have a similarly struc-tured day. Both networks typically start the day with several hours of news and current affairs followed by a mid-morning mix of more general programmes. There's lunchtime news followed by another programme sequence; then news again to coincide with the journey home for many commuters, and a further sequence of features and entertainment during the evening, ending with a late-night news bulletin. Specialist programmes such as science tend to get scheduled at times when the number of listeners tends to be lower, but independent Radio Joint Audience Research (RAJAR)[7] audience research shows that those who do listen seem to appreciate the programmes more (see also Marketing, Communications and Audience 2004).

Sources, stories and scoops

Twenty-five years ago, most science news stories arrived on paper, either through the post or via the teleprinter. There would be press releases, scientific journals, trade magazines and stories from news agencies torn off the teleprinters in the newsroom and forwarded on a rather random basis by the copy tasters there. On a weekday morning, the editor's in-tray was often half a metre deep in paper, even after a secretary had taken things out of their envelopes. As gatekeepers of this information, we got used to making a preliminary assessment of the newsworthiness of the information at a glance and the paper recycling bins soon filled up, as do their electronic equivalents today.

In those days, almost our entire output, both for programmes and scripts, was of short, semi-topical stories about newly published scientific research.[8] The first selection criterion was that they had to be important and significant to the wider public, not just to a few specialists in that field. Secondly, there had to be a story there that could be explained and made relevant to an often interested audience who were likely to be hearing about these issues for the first time. It's no good trying to report an important but incremental advance if it takes 20 minutes of background before anyone can understand why it is important. Thirdly, in most cases, there needed to be someone available to interview. Unlike print journalists, radio producers need an audio interview, preferably of good-quality, and that ideally means a face-to-face interview.

In the pre-digital era that biased us towards researchers located in London or at least within a day's return travel of London. However, as technologies improved, we found that we could also use studios in BBC local radio stations in many of the principal university cities. They had permanent lines back to London. We tried hard to make our coverage international, but that normally forced us to accept telephone quality for inter-views. Fortunately, many leading researchers came to London for conferences and very occasionally we got the chance to travel ourselves, in which case we would record almost anyone we could, feeding them into programmes over the next few weeks or months—all fine as long as topicality wasn't of prime concern.

Back at the daily editorial in-tray, the search for top stories becomes almost instinctive. The first subconscious selection was probably based on the nature of the source. An embargoed press release from a major scientific journal such as *Nature* or *Science* merited much closer attention than a release from a company or public relations (PR) agency, as did releases from universities and research institutes. One can quickly reject stories about staff appointments, grants awarded, buildings opened and minor prizes won. I can never

---

7. http://www.rajar.co.uk/.

8. For a further discussion of the criteria for selecting science news see Allan (2009).

quite understand why these get sent out in the first place—perhaps they are useful to a few local or specialist publications or at least to bolster the ego of vice-chancellors.

Inherent in the word 'broadcast' is the implication that the audience is wide and general. Therefore, there is no point in following up what I call second-order stories. These are stories that only apply to a tiny subset of the population, so a drug to treat the side-effects of an existing treatment for a rare disease would get left for the specialist medical press. We are not interested in the Mark 7 model of a product, even though it may have superior trim and redesigned flange gussets. On the other hand, some stories that are of no direct practical use whatsoever to our listeners are nevertheless amongst the best we cover. These might include interplanetary space missions or discoveries in particle physics, astronomy and molecular biology, as well as stories that are simply fun.

The news empire

The last 30 years has seen a big increase in the quantity and often the quality of broadcast news. At the start of that period, news and current affairs were regarded as specific programmes. Today, many news networks, and BBC World Service in particular, run a continuous 24/7 sequence of rolling news and current affairs and the remaining programmes are often seen, at least by news chiefs, as local opt-outs from that core sequence at certain times of day or night.

Throughout the 1980s and 1990s, the World Service Science Unit was steadily building its contribution to news. BBC audience research revealed that, in many parts of the world, science, technology and health—even if assessed separately—commanded high levels of interest and were often rated above arts and culture, popular music and even sport (for example see the BBC's International Broadcasting Audience Research (IBAR) internal reports for Thailand (1994) and Ghana (2005); see also Mytton and Forrester 1988). Many of the BBC World Service language services with limited air time were increasing their news and current affairs programming at the expense of the specialist programme slots that, in the past, had translated our scripts. They were taking their news agenda from the World Service newsroom and they in turn had a growing appetite for science. So members of the science unit were increasingly called upon to act as science correspondents, filing stories and providing expert analysis.

The increase in the proportion of news and current affairs in the schedules led to a slight reduction in the number of science programmes. At the same time, we wanted to make our programmes more distinctive. So, instead of turning out large numbers of low-cost magazine programmes, we switched to more crafted features and landmark series. Inevitably, they are more expensive to make. Hence, the channel controllers wanted to be more involved in selecting them. Their commissioning process can take a year or more. The result of all this is a polarization between the highly topical news stories of the day and carefully planned feature series that are less reactive to events.[9]

9. To listen again to an example of a feature series see *Discovery*,
   http://www.bbc.co.uk/worldservice/programmes/discovery.shtml.

A word about topicality

The rise of science on the news agenda has inevitably brought changes to the sort of story covered. In the 24/7 rolling context, science news prides itself on topicality, preferably to the hour or at least the day. Research announced or published at some non-specific time 'this month' is very unlikely to be considered sufficiently topical. If a story is deemed topical and important, news likes to revisit it frequently with every new development, but individual reports are short, perhaps only a minute or so in duration. That makes some complex research difficult to report at all, even if it is important, simply because there is not time to explain its significance. In my experience, this has resulted in a shift away from pure science reporting based on publications or conferences towards more controversial issues such as the spread of global disease, climate change and other topics which, with irresponsible reporting, could easily become 'scare' stories. In the rush for topicality this can lead to errors. Of course we hope it would not happen at the BBC but such are the news pressures at some organizations to be first with a story, regardless of accuracy, that it is possible to imagine them using the slogan 'we're never wrong for long'.

There are times when the science itself takes a while to get resolved. Sometimes this happens in virtual secrecy during the process of peer review (see Wager, Chapter 4.1 this volume). Sometimes the washing is laundered in public (see Thomas, Chapter 4.2 this volume for some examples). And sometimes it's interesting to report on that process. A good example is when a new near-Earth asteroid is first spotted. Often, the original calculations of its orbit have huge error bars, giving the possibility that it might collide with the Earth in a few years, decades or centuries. Journalists and sometimes scientists are often chided for reporting such objects before their orbits are better defined. But this is science as it happens, with all its attendant uncertainties and contingencies, and we should share it. The difficulty comes when we don't explain—or the headline writer fails to understand—statistics. On one hand it is like shouting 'Wolf, Wolf' and risks a real threat of getting ignored. On the other hand it brings issues such as asteroid impacts to the public's attention, which must be good as long as it is done responsibly.

Journalists, even science journalists, and especially news journalists, tend to hunt as a pack. Whilst we all relish the idea of an exclusive scoop, it only becomes a genuine scoop if it's a very significant story indeed. My first such scoop was when a story revealing the cause of AIDS was leaked to me in 1984. Another example was when my now-retired television colleague James Wilkinson broke the story about possible nano-fossil evidence for ancient life in a Martian meteorite (for discussion of the reporting of this issue, see Holliman 1999). With more run-of-the-mill stories, they need critical mass to become news. That typically means coverage in major daily newspapers and other media. No correspondent wants a call from an angry editor asking why they have not covered the story which is all over other outlets. So it is a common sight at press conferences to see science journalists from competing titles discussing with each other what angle they will be taking (see Allan 2009 for further discussion). I know of one science correspondent for a major newspaper who prefers to write three or four stories a day even if half of them do not run rather than have his editor accuse him of missing important science news. And timeliness can be everything. Several years ago I picked up an interesting astronomy story at a conference and broadcast a report on it. Three months later the same story was published in a major journal and made it to the front page of a couple of newspapers. I was asked by a news editor why I had not covered it, so I pointed out that I had done so 3 months earlier. He replied 'but it was not news then, it is now. Please cover it again'.

This herd tendency means that it is relatively easy to predict where many of the week's science stories will originate. Heading the list are the embargoed press releases of major scientific journals such as *Nature* and *Science*. Such releases are regularly e-mailed to selected journalists who promise to abide by the embargo, up to a week ahead of publication. There are pros and cons to this situation. Some, notably former BBC correspondent David Whitehouse (2007), say that it is a cosy system encouraging lazy journalism and boosting the power and the sales of the journals. I tend to disagree. By and large, scientists no longer leap out of their bath and run down the street shouting 'Eureka!' at the moment of a new discovery. The research tends to be incremental and often needs checking and refereeing before it is widely accepted. To become news it needs a topical peg and the critical mass of widespread coverage. Publication can provide that, and the embargoed release gives journalists a head start in producing high-quality coverage. Without the embargo it would be a mad rush. Everyone would want an interview immediately and there would be no time for careful writing, editing and checking or expert comment. On the other hand it does give perhaps undue power to the journals and makes scientists often reluctant to discuss their research before it is published.

Black and white, colour and shades of grey

Half a century ago, William Laurence was one of the few science journalists on the staff of the *New York Times*. In his opinion:

True descendants of Prometheus, science writers take the fire from the scientific Olympus, the laboratories and the universities, and bring it down to the people.

Nelkin (1995, p. 83)

There are still a few who share that view, particularly among academics promoting the public understanding of science (for example, see Durodié 2003, who argues that discussion of science should be left to experts). I would disagree: there is nothing divine about scientific research and the scientists themselves are seldom godlike. Some scientists, particularly those who sit on peer review boards and the editors of refereed journals, seem to prefer journalists to wait patiently until something is published and thereafter treat it as absolute truth. For me, half the fun is in the discussion, argument and controversy that precedes publication. Political journalists are not expected to keep quiet about a policy until it is written into the statute books, and nor should we. It can be argued that science should be falsifiable and based on evidence whereas politics often grows from opinion, but the true frontiers of science are always reaching into the unknown and there is much to be learnt from controversial theories even if they are ultimately shown to be wrong.

As journalists we have a responsibility to be accurate and unbiased. That responsibility is to our listeners and readers rather than to the holders of a particular body of knowledge.

We are no more godlike than scientists (considerably less, some would say). So sometimes we report accurately and fairly on something which later turns out to be false. I don't think we should be vilified for that, any more than should the scientists who were later proved wrong (so long as there is no deliberate deception involved). Therefore, I think we were right to report extensively, for example, the claims of Pons and Fleischmann in 1989 that they had created nuclear fusion at room temperature. In the end, those claims for cold fusion did not hold up but it was a lively and entertaining debate and, I think, also educational. It got the processes of scientific investigation and verification, and even electrochemistry, into the headlines and onto the front pages (for discussion of the reporting of this issue, see Lewenstein 1995).

There have been many other examples of controversial science and some of them are still not resolved. Some of the most engaging radio programmes that I have made concerned issues where science almost borders on theology; issues such as the nature of consciousness and whether there is any distinction between the mind and the brain, issues in physics and cosmology such as the role of the observer in a quantum system and whether we can talk meaningfully about universes beyond our own that we cannot even in theory ever observe.

Fascinating as such controversies can be, lines do sometimes need to be drawn. There are those who will push bad, inaccurate or even false science in order to promote their beliefs. Sometimes their arguments are easy to state and complex to refute, even though they may, ultimately, be wrong. There are dangers in even beginning to tackle such subjects, just as there are risks in entering into correspondence with the sort of listener who writes in with 20 pages of close handwriting in green ink. But there are fundamentalists on both sides. There are materialist scientists for whom one view of science has become so like a belief system that they refuse even to look at evidence that seems to counter it.

An obvious example of where it is easy to stray beyond the boundaries of 'science' is the debate between evolutionary biologists and religious creationists (see Allgaier and Holliman 2006 for a discussion of how this complex controversy was reported in relation to school science). It has become a very big issue in parts of the world, notably the USA where, for many people, belief seems to trump evidence. Occasionally, it comes to a head in the courts, usually over whether creationist beliefs should be taught in school biology classes alongside Darwinian evolution. (I suspect that if the USA had religious education classes, the creationist view could be included there and the argument would not arise, but the USA is supposedly a secular nation and therefore government schools do not teach religion.) The question is whether, as science broadcasters, we should discuss creationism in our programmes at all. Hopefully, the creationists have stopped demanding right of reply each time we talk about evolution in a science programme. Some strong-minded Darwinists such as Professor Richard Dawkins in Oxford won't even enter into discussions with creationists, but should science radio producers? News programmes rightly report major events such as court trials. I was recently asked to make a pair of programmes about the creation/evolution debate. I might not have done so in a science programme but this was for a religious slot and I was interested in the beliefs and assumptions of both groups.[10]

10. You can 'listen again' to the results at http://www.templeton-cambridge.org/fellows/redfern/
    publications/2007.03-04/heart_and_soul/.

Making programmes in the information age

Digital technology has transformed the way that media professionals work and the way that work is delivered. It seems hard to believe, but prior to 1986 we relied on postal services. After a brief affair with the fax machine, in the early 1990s the BBC World Service Science Unit acquired a primitive desktop computer with a telephone modem and we began to use e-mail. Incredibly, it was only in 1992 that the then Director-General, John Birt, visited our office to 'see the internet'. Using that slow telephone modem, we showed him several web pages, including a Canadian one for 'Birt's Used Cars'. As a result of that we were given two networked computers and funding to set up a pilot for a World Service web site. Today, the BBC web site has more than 34 million unique users—half of them in the UK—and receives around 600 million page impressions every month. As media professionals we now receive as well as give; the paper tower of the editorial in-tray has been replaced by hundreds of e-mails every day (see Trench 2009; see also Holliman 2007 for a discussion of how this is changing science newsroom practices).

Selecting stories from such wealth sometimes seems like trying to fill a drinking glass from the Niagara Falls. Releases from a few regular and reliable providers will get read. For many more, if they don't sell the story in the subject line they won't even be opened. For any press officers reading this chapter, don't ever begin 'we are pleased to send you this release from our company' or worse still, 'please see the attached press release'. It's fine to attach images and other information, or better still links to keep the e-mail size small, but if the essence of the story isn't there in text in the title, or at least the first paragraph, you might as well not bother. There are so many list servers sending out 'second order' stories that many of us don't even bother to subscribe to them. Even the better specialist science news digests (such as AlphaGalileo in Europe, EurekAlert! in the USA and SciDev.Net, covering international development), send so many stories that I don't always read right through even the weekly digests of titles (see Trench 2009 for a discussion of how these web sites are influencing science newsgathering practices).

Apart from giving us a deluge of stories without the need to fell forests, the internet is a fantastic tool for research—as long as it is used with critical caution. When reporting scientific research, we always like to track down the original researchers rather than relying on second-hand reports and pundits. For those of us without comprehensive contact books, search engines will find contact details for all but the most reclusive scientists. Very often you can find other papers and review articles by them or about them as well, though a frustrating number of journals require a subscription login before you can read their papers (see Gartner, Chapter 3.2 this volume for discussion of access to the contemporary scientific literature). The internet is also a quick way to check facts and research context. But a good journalist should never rely on a single source when producing a news story, whether for radio or any other medium.

Once you've found your story and made contact with the researcher, the next thing to do is to get a good quality audio interview, and here digital technologies make things much easier. Many cities around the world now have radio studios equipped with an Integrated Services Digital Network (ISDN) connection. There is usually a charge for using such studios, but it is a lot less than the cost of sending a reporter out there.

The result is a linkup in good quality and real-time either to pre-record or even broadcast live. Another option that can be cheaper and easier is if the interviewee has access to a reasonable digital recorder, the broadcaster can ask questions over the phone and they can record their answers locally and then send the audio file using a system such as file transfer protocol (FTP). It's not instant but it is a lot quicker than express-mailing an analogue tape recording.

Assuming your show is not live, once you have gathered all your interviews and location recordings, the next thing is to edit them. Not so long ago that used to begin with dubbing them onto open reel recording tape and then physically cutting the tape and joining it with sticky tape. Now you just transfer your digital recording at high speed onto your computer or laptop and use one of the many excellent digital editing packages to craft your programme, often mixing in music, sound effects and narration without even needing to visit a studio. 'Ums' and 'ers', features of everyday speech, disappear at the click of a mouse and your biggest problem is squeezing all that excellent material into what is often a meagre transmission slot.

Texture and colour in sound

Good radio is far more than a lot of talk. As one listener is supposed to have remarked, 'I much prefer drama on radio to television; the scenery is so much better'. There are many ways to build scenery out of sound. Central to all of them is high technical quality. Now that almost anyone can make multi-platform audio programming, for example for the web or podcasts, quality is one thing that separates professional broadcasters from the rest. Wherever possible we try to use the best equipment, rooms with good acoustics and, for remote interviews, high-quality lines rather than telephone. To the trained professional, it is amazing how some podcasts sound as if they were recorded in bathrooms.

Of course, if the story is about bathrooms, then suddenly that acoustic becomes relevant and even desirable. The same is true of background noise. Traffic, chattering people, noisy machinery and the rest are distracting if they are nothing to do with the story or its setting. But the best interviews are done on location, and there relevant noises can help you paint a picture in sound. But again sound quality is important. It's no good if you can't hear what contributors are saying over the noise and it's not much help if you're given no clue about what the noise is and why it's there.

Listening in the information age

The days when the BBC had a virtual monopoly on quality speech broadcasting are long gone. In most parts of the world today there is a bewildering array of radio stations providing an endless stream of local news, chat and music in every style imaginable. Those providing comprehensive, accurate international news, and in particular science, are in shorter supply but this broadcasting landscape is still highly competitive. And now we are beginning to see an unprecedented change in the way people listen: instead of tuning to their favourite station schedule to hear whatever happens to be being broadcast at the time, many people are seeking out particular programmes to 'listen on demand' over the internet, download onto their portable players or subscribe to podcasts.

The BBC has been offering audio on demand for several years and it's proving very popular. In November 2007 there were about 20 million on demand programme requests across the UK networks, a figure that is rising steadily (see the section on 'Useful web sites' at the end of the chapter for examples). A few programmes such as *Digital Planet* on the BBC's World Service and *In Our Time* on BBC Radio 4 have been podcast as download-able MP3 files for several years. At the time of writing (2008) many other science speech programmes have been added in just the last few months and download numbers are still rising rapidly. For example, during October 2007 there were 159,398 downloads of the *Digital Planet* podcast, 59,243 of *Science in Action* and 45,315 of *Material World*—not bad for weekly programmes that had only been podcast for a month.

The other side of that upsurge in digital listening is that anyone with a little basic equipment can create podcasts. Of course, amateur radio enthusiasts are as old as radio, but they have not been seen on this scale before. Suddenly, professional radio pro-ducers have competition not only from established rival broadcasters but also from the publishers of newspapers and magazines, science organizations and the aforementioned 'pro-ams' (Editorial 2006). Rather than me adding a three-page appendix to this chapter, just put 'science podcast' into the internet search engine Google[11] and you'll see what I mean. Some are better than others, but it is clear that the information-hungry will never 'starve' again.

In this competitive world, it is no good just making the programmes you want to, or giving listeners what you think they ought to hear, or worse still what scientists think they ought to hear. The BBC continues to be the best-known global brand name in the business, certainly in terms of public sector broadcasting, still commanding a great deal of trust, respect and loyalty. But even BBC radio producers have to give listeners what they want or they will switch off. But I do not think that necessarily means 'dumbing down'. Quite the opposite, it means making programmes that are both difficult and exciting to make and stimulating, informative and sometimes surprising to listen to.

■ REFERENCES

Allan, S. (2009). Making science newsworthy: exploring the conventions of science journalism. In: *Investigating Science Communication in the Information Age: Implications for Public Engagement and Popular Media* (ed. R. Holliman, E. Whitelegg, E. Scanlon, S. Smidt and J. Thomas). Oxford University Press, Oxford.

Allgaier, J. and Holliman, R. (2006). The emergence of the controversy around the theory of evolution and creationism in UK newspaper reports. *Curriculum Journal*, **17**(3), 263–79.

11. http://www.google.co.uk.

Durodié, B. (2003). Limitations of public dialogue in science and the rise of the 'new experts'. *Critical Review of International Social and Political Philosophy*, 6(4), 82–92.

Editorial (2006). Sound science. *Nature*, **439**, 2.

Eldridge, J., Kitzinger, J. and Williams, K. (1997). *The Mass Media and Power in Modern Britain*. Oxford University Press, Oxford.

Franklin, B. (1997). *Newszak and News Media*. Arnold, London.

Holliman, R. (1999). Public affairs media and the coverage of 'life on Mars?'. In: *Communicating Science: Contexts and Channels* (ed. E. Scanlon, E. Whitelegg and S. Yates), pp. 270–86. Routledge, London.

Holliman, R. (2007). Reporting environmental news: the evolving context for newspapers in the digital age. *Frontiers in Ecology and the Environment*, 5(5), 277–8.

Holliman, R. (2008). Communicating science in the 'digital age': issues and prospects for public engagement. In: *Readings for Technical Communication* (ed. J. MacLennan), pp. 68–76. Oxford University Press, Toronto.

Leadbetter, C. and Miller, P. (2004). *The Pro-am Revolution: How Enthusiasts are Changing Our Economy and Society*. Demos, London.

Lewenstein, B.V. (1995). From fax to facts – communication in the cold-fusion saga. *Social Studies of Science*, **25**(3), 403–36.

Marketing, Communications and Audience (2004). *Science on the Radio*, November. (Internal BBC Report.)

Mytton, G. and Forrester, C. (1988). Audiences for international radio broadcasts. *European Journal of Communication*, **3**(4), 457–81.

Nelkin, D. (1995). *Selling Science: How the Press Covers Science and Technology*, 2nd revised edn. W.H. Freeman, New York.

Trench, B. (2009). Science reporting in the electronic embrace of the internet. In: *Investigating Science Communication in the Information Age: Implications for Public Engagement and Popular Media* (ed. R. Holliman, E. Whitelegg, E. Scanlon, S. Smidt and J. Thomas). Oxford University Press, Oxford.

Whitehouse, D. (2007). Science reporting's dark secret. *The Independent: Media Weekly*, 23 July, p. 13.

■ FURTHER READING

- Hendy, D. (2000). *Radio in the Digital Age*. Polity, Cambridge. Although this book does not deal specifically with science on the radio, it does consider it in a global context, examining issues of production and reception, the role of the internet and the political economy of radio.

- Lewis, P. and Booth, J. (1989). *The Invisible Medium: Public, Commercial and Community Radio*. Macmillan, Basingstoke. Again, this book does not deal specifically with science on the radio. Rather, it provides an historical account of the development of public sector broadcasting and commercial radio with a focus on the UK in particular, but also examining America, Canada, Australia, parts of continental Western Europe and community-based radio.

- Mazzonetto, M., Merzagora, M. and Tola, E. (2006). *Science in Radio Broadcasting: the Role of Radio in Science Communication*. Polimetrica, Milan. One of the few books to specifically address science on the radio, this book is mainly aimed at media professionals working with this form of science communication.

■ **USEFUL WEB SITES**

- BBC Radio 4 Science: http://www.bbc.co.uk/radio4/science/. This web site lists the range of science programmes now archived and available as .ram files.

- BBC Science/Nature News: http://news.bbc.co.uk/1/hi/sci/tech/default.stm. This web site is regularly updated with the latest science news stories.

- *The Guardian* science podcast: http://www.guardian.co.uk/science/podcast. This is a weekly podcast produced by *The Guardian* newspaper, covering science, technology and environmental news stories. Previous podcasts are archived on the site, and there is an associated blog.

- The Naked Scientists: http://www.thenakedscientists.com/. The Naked Scientists team are described on their web site as 'a media-savvy group of physicians and researchers from Cambridge University who use radio, live lectures, and the Internet to strip science down to its bare essentials, and promote it to the general public'.

SECTION 6

Practising public engagement

. . . once we have science talked about—and people thinking about it and in a sense owning it—then we have a chance.

Susan Greenfield (2004) quoted in *The Guardian*

6.1 **The development of *Our Dynamic Earth*,** by Stuart Monro   195

6.2 **Engaging through dialogue: international experiences of Café Scientifique,** *by Ann Grand*   209

6.1

The development of *Our Dynamic Earth*

Stuart Monro

The background to *Our Dynamic Earth*

Edinburgh is a city of contrasts. The castle sitting high on its basaltic rock plug dominates the city. The Royal Mile extends eastwards down to the Palace of Holyroodhouse and is known geologically as a crag and tail ridge, moulded by the movement of ice during the last glaciation. On either side of the Royal Mile is the Old Town of Edinburgh, a World Heritage site which is the cornerstone of Edinburgh's tourism industry. But go back to 1777 and this historic site of Holyrood, adjacent to the palace, was also the heart of the brewing industry. The firm William Younger and Company, later to become Scottish and Newcastle Brewers Ltd, made use of the abundant supply of subsurface water to develop many well-known brands of beer, until in 1986 the decision was made to concentrate brewing activity at the Fountain Brewery in the west side of the city. This left a legacy of industrial dereliction in an important historic part of Edinburgh, adjacent to the World Heritage site.

Sir Alex Rankin, then Chairman of Scottish and Newcastle Brewers Ltd, rather than placing the site on the open market, urged his Board to establish a foundation to provide support for a project worthy of the location which, at its heart, would have a major new landmark building for the City of Edinburgh. This would house a visitor attraction with a strong educational dimension, promoting an understanding and appreciation of the natural world. This concept is what is now regarded as a 'science centre' though there are still differing opinions as to whether or not this is an appropriate descriptor for what is now *Our Dynamic Earth*. Adjacent to the land to be disposed of by Scottish and Newcastle Brewers was also the site of two gasometers owned by British Gas. This small but significant area of land, contaminated with all the infrastructure of the gasometer installation, was also incorporated within the redevelopment plan. The Holyrood Brewery Foundation was tasked with overseeing the transfer of the land, including that donated by British Gas, to achieve the regeneration objective.

The Foundation drew on a wide range of expertise to consider how the area should be developed; it was of direct interest to the City of Edinburgh Council and its tourism development strategy, it was of concern to the Edinburgh Old Town Renewal Trust because of their interest in Old Town architecture and it was important to neighbours in the Palace of Holyroodhouse and in the Dumbiedykes flats because of the changes in the local environment. The landmark building to house *Our Dynamic Earth* was already agreed as an integral part of the project but the development of the other areas which were part of the land donation had to be treated with great sensitivity. The area, lying at the heart of the Old Town of Edinburgh, had to have a development plan which would place new 21st century architecture adjacent to the 17th century architecture of the Old Town and would provide the diversity of buildings that would bring new life to the Holyrood area.

It was recognized that the regeneration of the Holyrood area could not be achieved by merely parachuting a visitor attraction into the area. For Holyrood to become a destination for visitors to the city there needed to be a critical mass of other activity. There was a requirement to have people living within the area, so housing became a feature. There was a requirement to bring people in to work, so attractive office accommodation also became a feature, and to support these and the visitors to the area there was a need for cafés, shops and restaurants. An architectural master plan emerged which brought all these elements together and welded them into a plan which was appropriate for the location. The outcome has been that a visitor can walk from the 17th century Canongate Kirk, through Bakehouse Close and follow a changing architectural style to modern buildings fit for the 21st century that maintain something of the stylistic legacy of Old Edinburgh.

Roots of *Our Dynamic Earth*

An important aspect of the establishment of this visitor attraction was that it should have its roots firmly grounded in the local culture. In considering what subject matter should be covered, it was important to look at the geographical setting, the features in the surrounding landscape and the legacy of scientific thinking that was part of the culture in this part of Edinburgh.

Plans had been drawn up for a living tropical rainforest on the site, but the costs of building and, more importantly, maintaining such a concept were prohibitively high. The concept of the Younger Universe emerged, linked to the family name of Younger who were associated with the original brewery. But it was only when the landscape surrounding the location and Edinburgh's historic legacy as a centre for the development of scientific thinking were considered that it became obvious that the story to be told should be about the processes that formed our planet and the rich diversity of environments that it contains; a planet which has undergone continual change now and into the future; a 'Dynamic Earth'.

Unlike other scientists, those that are concerned with the natural world have the complete environment as a laboratory. Examples of this abound. The city of Edinburgh is surrounded by a landscape that has been fashioned by the powerful forces of the Earth. The global processes of plate tectonics placed Edinburgh in the tropical regions between

400 and 300 million years ago and the sedimentary rocks formed at that time contain evidence of shallow tropical seas and hot humid coal forests. The colder conditions during the Ice Ages of the last 2 million years produced thick ice sheets, which shaped the surface and contributed to the rugged and diverse landscape in which Edinburgh is situated. In the heart of the city, the extinct volcano of Arthur's Seat is an example of a natural feature that can be enjoyed for its own sake yet also provides an opportunity to understand the processes of volcanism. Edinburgh is therefore an illustration of global processes that have been operative through geological time.

The resources that support the community have also been sought from the rocks in and around the city. Salisbury Crags and other igneous sills in the city at Corstorphine Hill, Turnhouse Hill, Mons Hill and Dalmahoy Hill have all been quarried at some time to provide building or road stone. Quarrying at Salisbury Crags, however, was proceeding so rapidly in the mid-18th century that the skyline of Edinburgh was being affected. Some of the original sections were quarried away and it is fortuitous that extraction ceased when it did, maintaining the characteristic profile of the Crags and important exposures at their base. The sandstones of Edinburgh, most notably at Hailes and Craigleith, have been extensively quarried to provide the stone for many of Edinburgh's most important buildings and have also been exported to London and elsewhere. Water from the thick glacial deposits immediately adjacent to Arthur's Seat was the basis for the brewing industry of Holyrood. Energy resources, originally obtained from the coal deposits around Edinburgh, are now entirely mined by opencast methods. Today, oil and gas from the North Sea provide for much of Scotland's energy needs. The limestone and mudstone of East Lothian form the basis of a cement industry that underpins much of the civil engineering activity in Scotland.

The resources of the Earth remain an essential component in the fabric of modern society but many are finite and need to be utilized in a sustainable way to ensure their longer-term availability. Decisions on the utilization of resources require that earth science information is integrated with planning, economic and social science data as part of the strategic planning process. These are environmental issues that impinge on everyone. As citizens in the 21st century we are all being required to have an informed opinion on issues of the day, many of which are underpinned by an understanding of how the Earth works. This understanding will come through a variety of media, some more authoritative than others. Newspapers and television will offer some of that understanding; however, it needs to be provided in an engaging way, synthesized so that key messages are communicated effectively. This is becoming the role of science centres throughout the world, and it is therefore one of the objectives of *Our Dynamic Earth* to give visitors of all ages, the citizens of today and those of tomorrow, an insight into the science that underpins environmental issues so that they may make informed judgements.

Edinburgh is an appropriate site for *Our Dynamic Earth* because it is a superb example of how geological forces, in the form of fire, ice and water, can shape a landscape and give character to a city. The evolution of Edinburgh as a beautiful scenic town was a conscious exercise by the city elders in which the physical surroundings played an integral part. Equally, however, Scotland itself is an ideal location for such an exhibition as, within its limited compass, there is an unparalleled range of geological history, from some of the oldest rocks in the world to some of the youngest, post-glacial deposits.

The importance of James Hutton

The development of the Earth Sciences is firmly embedded in the cultural history of Edinburgh in the person of James Hutton. Hutton was an Edinburgh man through and through, educated at the High School and Edinburgh University. His father, who died when Hutton was still young, was a wealthy merchant and Hutton inherited his business acumen. He was involved in a number of commercial enterprises, including the construction of the Forth and Clyde Canal, and owned considerable property, some of which would have fallen within the area of the present urban regeneration project. Therefore, the location of *Our Dynamic Earth* is singularly fitting. It looks out on part of the evidence that inspired Hutton's *Theory of the Earth* (1788) and lies close to the home on St John's Hill where he set out his ideas on paper.

Hutton was a genius, and his perception of the heat engine of the Earth and the resultant processes led to his title 'the founder of modern geology'. Perhaps more significantly, in casting off the shackles of theology in measuring the age of the Earth, he demonstrated freedom of thought regardless of orthodoxy. It is worth recalling that more than half a century later Hugh Miller entitled his book *The Testimony of the Rocks; or, Geology in its Bearings on the Two Theologies, Natural and Revealed* (1857). The revolutionary significance of Hutton's concept of time was wittily highlighted by Louis Agassiz in his Memoir to Miller's earlier book *Footprints of the Creator* (1849). Here he recalled the words of Sir David Brewster:

The Geological Society of London, which doubtless sprung from the excitement in the Scottish metropolis, entered on the new field of research with a faltering step. The prejudices of the English mind had been marshalled with illiberal violence against the Huttonian doctrines. Infidelity and atheism were charged against their supporters; and had there been a Protestant Inquisition in England at that period of general political excitement, the geologists of the north would have been immured in its deepest dungeon.

(Agassiz, 1849)

But Hutton was endowed with an enthusiasm and an insatiable curiosity about science which did not confine his interest to geology but extended it into all aspects of the living world. He had a holistic view of the Earth. The conclusion of the abstract of Hutton's *Concerning the System of the Earth, its Duration and Stability* (1785) reads:

. . . there is opened to our view a subject interesting to [a] man who thinks; a subject on which to reason with relation to the system of nature; and one which may afford the human mind both information and entertainment.

This statement chimes exactly with the aims and objectives of *Our Dynamic Earth*, which is designed to be both a visitor attraction and an educational resource.

Funding of *Our Dynamic Earth*—the capital project

The total cost of the building and exhibition construction, associated infrastructure and site preparation works, together with the establishment of the operating company was £40 million. Funding came from two main sources. Part of the funding for *Our Dynamic*

Earth was provided from the sale of land on the two adjoining sites for commercial development, land gifted specifically for this purpose. The architectural master plan was used to enforce a design template for coordination of the whole area, ensuring a variety and quality of style and the continuation of a recognizable 'Old Town' urban character. The site on the south side of Holyrood Road initially required extensive decontamination work following the inheritance from British Gas, but provided the site for the headquarters office for *The Scotsman* newspaper.

The second main source of funding came from the Millennium Commission who awarded a grant of £18,070,933, allowing the project to go ahead on the basis that all the capital costs could be covered.

Development of *Our Dynamic Earth*—the science story

The vision and mission statement for *Our Dynamic Earth* as agreed in 2007 is what the scientific story seeks to deliver (Box 1).

BOX 1 THE VISION AND MISSION STATEMENT FOR *OUR DYNAMIC EARTH*

The vision

To excite and engage people of all ages towards understanding how the Earth works: this small blue planet hanging in the infinity of space is our only home.

The mission statement

To provide an outstanding and dynamic visitor experience presenting current and topical information promoting understanding of earth and environmental sciences in a fun and entertaining way.

The first step in developing the scientific story was writing an interpretative plan which embodied the key messages that should be communicated. This evolved over time, at each stage drilling deeper and deeper into the subject matter. At the highest level there was the desire to emphasize the dynamic nature of the planet to engage visitors ranging in age from young children to the elderly who were assumed to have little knowledge of the astronomy, cosmology, geology or environmental science associated with Earth's processes and environments. In presenting the dynamic nature of the planet, the story started by exploring the ways in which the Earth is currently being monitored; how we understand the dynamics of the Earth's system from satellite sources and that of the Universe in which we sit, set against a backdrop of geological time. This was followed by chapters exploring the main processes—geophysical, atmospheric, hydrological and biological—that have shaped the formation and development of the Earth and its diverse environments, culminating in an examination of anthropogenic influences on the planet.

The interpretative plan also sought to relate the scientific story to the national curricula of Scotland, England and Wales and Northern Ireland, creating a national (Scottish) and international educational resource. Earth and environmental science is embedded within the curricula throughout the UK and it was an objective of the interpretative plan to develop a resource which would be available to teachers and supported by other activities delivered by a dedicated educational team. It subsequently proved to be a valuable resource with 40,000 school children visiting in 2006/07 as school groups either to tour the exhibition or to engage in activities in *Our Dynamic Earth*'s 'Discovery' rooms.

The writing of the interpretative plan very quickly developed into telling a story of the development of the Universe, our own planet Earth, the physical and biological processes that shaped it and the rich diversity of environments that were produced. The story was fundamentally linear, and as a result the structure and layout conceived by the designers, Event Communications, was linear in form. This makes *Our Dynamic Earth* rather different from most conventional science centres. (Working with the designers to deliver an innovative experience is discussed later—see page 202–3.)

The scientific story of *Our Dynamic Earth*

Our Dynamic Earth is divided into the following areas:

- **'The State of the Earth'** is the first section of the exhibition and looks at the ways in which the Earth is being continually monitored to build up a picture of a dynamic planet. Here is the opportunity to provide information about what is happening in the world today, the latest earthquake or mission into space. It aims to raise many questions in the mind of the visitor about the processes that cause this continual change and demonstrates how satellite technology allows us to 'take the pulse' of the planet.

- **'The Time Machine'** could be regarded as just a lift taking you from one floor to another. However, suspending reality for a moment conveys the concept through visuals and movement—the speed of the lift slowly increases to illustrate the immensity of geological time in contrast to recorded human history—taking the visitor from their present familiar surroundings, through the World Wars, the Roman invasion, the Stone Age, human origins, the age of dinosaurs, early life forms, to a planet devoid of life and finally to the beginnings of the universe in the Big Bang—where the story begins.

- **'How it all Started'** takes the visitor from the Big Bang and the start of the physical universe to a planet capable of sustaining life. It demonstrates the relative place of the Earth within the Universe and the vast distances involved. The environment is the deck of a spaceship looking out at images of the building blocks of the Universe from early galaxies to our own solar system and tiny planet.

- **'The Restless Earth'** is concerned with the internal processes on this planet: what causes volcanoes and earthquakes, how mountains are built up and why continents move. Here, amid a volcanic terrain with smoke and heat, the visitor can experience the products of a mid-ocean ridge volcano and the pyroclastic flows associated with volcanic eruptions where one tectonic plate moves below another.

- **'Shaping the Surface'** examines the surface processes sculpting the landscape, particularly glaciation, which has been an important agent in moulding the scenery of Scotland. Here in a chilly environment you can take a helicopter journey across a modern

Norwegian glacier seeing how the landscape is formed and comparing that with the style of landscape in Scotland today.

• **'Casualties and Survivors'** looks at how biological processes have influenced the planet. The evolution of the skeleton left fossilized evidence of a wide range of life forms. Some of these still have living relatives: these are the survivors. Others are only preserved as fossils and are the casualties. Mass extinctions have taken place in the past and will undoubtedly be a feature of the future. In this context the exhibition asks will *Homo sapiens*, 'the human animal', be a casualty or a survivor?

• **'The Environmental Dynamic'** is the mid-point of the exhibition and illustrates through large-scale satellite images the impact that people have on the landscape for both good and ill. 'Man is no longer a figure in the landscape; he is a shaper of the landscape' (Bronowski 1973).

• The exploration of **'The Oceans'** has been moved on immeasurably by remote sensing of the ocean floor. The use of satellite technology has greatly increased our understanding of the linkages between the oceans and the atmosphere and the transfer of momentum, energy and matter at the ocean–atmosphere interface. Here the oceanic environment can be explored with the sense of the ocean's movement conveyed by moving banners and the marine environments viewed through a 'yellow submarine'.

• **'The Polar Regions'** represent the extremes of cold climate, yet though the Arctic and the Antarctic are superficially very similar, one is oceanic and the other continental. Ice is a fundamental part of both the Arctic and the Antarctic and the area is dominated by a large ice sculpture in an environment which is distinctly chilly.

• **'The Journey of Contrasts'** portrays the range of environments from the tundra to the savanna grasslands and how they vary from day to night and through the seasons. The area also explores what is needed to support life: food, water, energy and shelter.

• In **'The Tropical Regions'** the visitor experiences the hot, wet environment of the tropical rain forest, where the Sun's energy encourages the greatest diversity of plant and animal life. Here not only is a rain forest from south-east Asia re-created but the structure of the forest is examined in relation to the animals that live in the different 'layers' of the forest. In each of these experiences, whether related to a process or environment, there has been a conscious attempt to create an experience which is immersive; stimulating all the senses to complement what is seen.

• The **'Showdome'** was initially intended to be a summary of the exhibition, highlighting the dramatic processes that have shaped the planet Earth, many of which may be regarded as 'hazards' but many of which, nevertheless, are also the processes of renewal which rejuvenate the planet. It was the conclusion that set these dramatic processes in the context of humanity that inhabit much of the planet.

Throughout the different areas the combination of text, images, sound, movement, temperature variations, etc. help to communicate to the visitor the different environments and geological processes that formed these by relating them to their own experiences and by using a range of senses, the aim being to widen the appeal of the exhibitions to a range of ages, differing levels of prior knowledge and mental and physical abilities.

The linear narrative of *Our Dynamic Earth*

The formulation of an interpretative plan highlighted how similar the exhibition was to a story. 'Once upon a time, a long time ago, there was a Big Bang . . .'. Most conventional science centres around the world operate on the principle that the visitor moves around the building in no particular order, engaging with different exhibits that demonstrate different aspects of science. The hope is that these often partial interactions will produce a better understanding of the science. I would question that assertion. While it is appropriate to challenge the visitor with the science being presented we have assumed that the typical visitor does not have a deep understanding of the Earth and environmental processes. I would argue that it helps the visitor to place that learning in a holistic context, to represent the scientific story in a sequential way with one chapter building on the previous one. Here the analogy with a storybook is very apt. Subsequent feedback from visitors demonstrates that this approach is successful in conveying the different elements of the *Our Dynamic Earth* story and how they relate to each other.

This approach does have inherent strengths but also weaknesses. The strengths are that it is very much easier to ensure that a coherent story is being told and the visitor emerges at the end of the exhibition with some ideas about how the processes that shaped the Earth work and of the rich diversity of environments formed by these processes— a quality learning experience. The weakness is that it is difficult to change in order to respond to new scientific developments, in response to new presentational styles and to react to demand for something new from repeat visitors. It is also true that the visitor also perceives it as a linear experience and having visited once, may conclude that *Our Dynamic Earth* is 'done and dusted'. Regular changes to the experience are therefore necessary to drive the repeat visit, particularly from those living in the local area in and around Edinburgh.

Dialogue with designers

Dialogue with designers is an important part of the process. The creation of an exhibition like *Our Dynamic Earth* requires the efforts of people with a wide range of skills. The scientist's answer to conveying a scientific concept or process might be to produce a diagram. The designer takes that idea and gives it added value by looking at more innovative ways of communicating. In *Our Dynamic Earth*, for example, we wanted to communicate the concept of subduction where one tectonic plate moves underneath another. I produced a diagram but the designer converted that into an audio-visual experience where the visitor takes a much accelerated journey riding the oceanic plate as it moves down a subduction zone beneath a continental plate with the resulting melting and production of violent volcanic eruptions and earthquakes; a truly immersive experience.

The design of the exhibitions at *Our Dynamic Earth* also differs in other ways from those found in a number of science centres. When *Our Dynamic Earth* was being designed, computer interactives were popular, but when I made a visit to a science centre in Holland my observations of how these were used caused me to question this approach. *New Metropolis* (now redesigned and rebranded as *Nemo*) is a science centre in Amsterdam that uses a variety of computer-based and hi-tech interactives to communicate aspects of science.

This style of delivery was observed to be very popular with children and no sooner had one child finished using an exhibit than another one was beginning. So what about the less forthright and usually older people, myself and my wife included, who were less aggressive in gaining access to these interactives? We didn't get a chance to interact with many of the exhibits and hence did not fully engage with the scientific ideas being communicated. Such interactives do have their place, but it should be recognized that they are almost exclusively a one-to-one activity. Group interactivity that more accurately represents how science is conducted in society should also have its place on the exhibition floor. The technologies to deliver such interaction are only now being developed but hold promise for the future.

Identification of the commercial imperative for longer-term sustainability

Though the initial, and perhaps ill-conceived, business plan for *Our Dynamic Earth* was that it would be self-funding, it became apparent as soon as the operating team were put in place that such a model could only be achieved if a proactive commercial stance was adopted. A science centre has a number of standard income streams that can be addressed in addition to the exhibition itself. A shop which has its own particular brand is a useful income generator, as is a café, food being an important facet of any day out. The challenge is in finding other ways of generating income. In the case of *Our Dynamic Earth*, it was recognized that the new Scottish Parliament would make the Holyrood area much more of a 'destination'. In addition, the regeneration of the area with the unique style of building made the potential of hosting a range of events in *Our Dynamic Earth* an interesting prospect. While corporate hospitality may be viewed simply as an exercise in income generation, it can also be seen as an 'outreach' activity. When guests visit *Our Dynamic Earth* for a corporate event some aspects of environmental science may rub off on them. Many corporate dinners include a reception in the galleries or a tour of the whole of *Our Dynamic Earth* as part of the event. An analysis of the corporate business in 2006/07 indicated that around 40% of the corporate events had a strong educational dimension. Some were to do with science communication such as the BA Scottish Science Communication Conference; others had more of an educational focus with *Our Dynamic Earth* being used as a venue for science-based continuing professional development (CPD) events for teachers and others.

Science centre funding and the creation of a Scottish Science Centre Network

The Millennium Commission funded capital projects but did not contribute to revenue funding. Throughout the world there are virtually no science centres that survive without some public funding. Revenue funding and funding for future developments therefore became increasingly important not only to *Our Dynamic Earth*, but to all the science centres in Scotland. One of these, *The Big Idea* was also funded by the Millennium Commission. It occupied the site of the former Nobel factory just to the north of Irvine in Ayrshire and opened on 15 April 2000 as an inventor centre stimulating creativity in young people especially. In 2003 *The Big Idea* went into receivership and closed its doors to the public. It was recognized that what had happened to *The Big Idea* could also happen to other science centres in Scotland in a 'domino' effect. Following on from this event the Chief Executive Officers of the remaining Scottish science centres were in discussion with Jim Wallace, then Deputy First Minister and minister with responsibility for science in the Scottish parliament, concerning future funding options.

Undoubtedly, the demise of *The Big Idea* was the catalyst for action to protect the sizeable Scottish asset which was represented by the four other science centres, namely *Satrosphere* in Aberdeen, *Sensation* in Dundee, *Our Dynamic Earth* in Edinburgh and the *Glasgow Science Centre*. Other initiatives were also under way, most importantly the work of the Scottish Science Advisory Committee which had been set up in the wake of the publication of the Scottish Science Strategy to advise ministers on science policy. The committee recognized the value of informal science education and felt that as museums and art galleries were funded from the public purse, organizations which communicated the science which underpinned the Scottish economy should likewise be the recipient of public money.[1] In June 2004, acknowledging the recommendations of the Scottish Science Advisory Committee, ministers decided to initially provide £5.1 million over 2 years for the four remaining science centres. Funding subsequently stabilized over the 2006/07 and 2007/08 periods at £3.7 million. The critical statement is that an investment of £3.7 million per year was protecting a £130 million investment and communicating science to over 700,000 people. As part of the funding package a 4-year strategy was developed by the Scottish Science Centre Network with the Scottish Executive which set objectives for the period 2005–09.[2]

The strategy outlines the objectives of the Scottish Science Centre Network in relation to:

- science curriculum 3–18;
- links with further and higher education institutions;
- links with industry;
- encouraging science as a career.

This is seen as an evolving strategy forming the basis of the development of the Scottish science centres as catalysts for science communication. Now that the Scottish Science Centre Network is functioning it is becoming clear that effective science communication will be very much better if connections can also be made with the SetPoints (see the useful web sites at the end of this chapter), the Science Festivals and other science communicators in Scotland. The objective is to allow access to a science event to at least all children in Scotland and by an integrated approach there is a greater chance of achieving this objective.

1. http://www.scottishscience.org.uk/main_files/pdf/Publications/final_Annual_Report.pdf.
2. http://www.scotland.gov.uk/Publications/2005/12/06113103/31038.

It is becoming increasingly important that the benefit of science centres in bringing added value to science engagement is demonstrated and quantified. In Scotland this has been done by Her Majesty's Inspectorate of Education (HMIe).[3] An inspection took place in November 2007 which resulted in the publication of a report highlighting the strengths and weaknesses of each of the science centres. The follow-up to this report has been the submission of action plans by each science centre to address all of the issues raised and a further action plan detailing the future activities of the network. This report has proved to be a valuable stimulus to improvement in the educational product in each centre. A further inspection of *Generation Science*, the 'outreach' arm of the Edinburgh International Science Festival, is currently (2008) being carried out.

Keeping *Our Dynamic Earth* refreshed

There are many drivers for refreshing the exhibits in *Our Dynamic Earth*. Notwithstanding some of the challenges already noted, the overall linear narrative of the exhibition demands continual change in order to attract repeat visitors and to address the evolving science and the changing presentational technologies. Visitor feedback, collected daily through questionnaires, is also important in defining areas for change as is feedback from the scientific community. In addressing the latter, the scientific advisory board that commented on the original development plans now has a role in advising on areas to update and change. The opportunities for change are also driven by the availability of funding and the advent of support for this through the ReDiscover Fund and through funding from the then Scottish Executive (Box 2). These were also the stimulus for a number of refreshment developments.

BOX 2 THE REDISCOVER FUND

ReDiscover was launched in July 2002, as a joint venture between the Millennium Commission, the Wellcome Trust and the Wolfson Foundation. The Commission invested £26 million in this £34 million scheme, which offered a unique opportunity for science centres and museums to renew and refresh their exhibits. ReDiscover allowed them to redevelop existing resources to ensure that exhibits were kept fresh and up-to-date, and visitors were given new experiences.

Our Dynamic Earth was successful in three out of the four bids it submitted to the ReDiscover Fund and this, together with matching funding from the then Scottish Executive, has kept it up-to-date with current developments in the earth and environmental sciences.

Feedback from visitors suggested that *Our Dynamic Earth* was not as interactive as it might be, especially for younger children. It was felt that there was an opportunity for more interactives that had a mechanical feel to them rather than those styled on computer games (for a discussion of the use of interactive exhibits and their effects on

---

3. http://www.hmie.gov.uk/documents/publication/sscn.pdf.

children see Meisner and Osborne 2009). The first successful bid to the ReDiscover Fund enabled 11 interactives to be introduced into the galleries, each carrying one of the key messages originally determined in the interpretative plan. While many of the interactives had a mechanical feel, involving turning handles and pressing buttons etc., there was a strong digital electronic component with animations playing an important part. Where possible the design of individual interactives was such as to allow more than one person to participate.

Ensuring that exhibits are successful in communicating the message is important, as they represent considerable investment. These interactives were tested by a group of children at an early stage in their development. Some were found to be less than effective at communicating, some required too much instruction, and others still worked well. This formative testing allowed changes to be made and ensured their success on installation.

Another aspect of feedback from visitors suggested that the external space was somewhat austere and indeed some visitors thought that the building was the new Scottish Parliament. In addition, there were further aspects of the story of *Our Dynamic Earth* that the Scientific Advisory Committee wished to tell. It was also becoming apparent that a science centre could benefit from both an external as well as an internal space. All of these aspects came together in a second bid in support of some external exhibits which would support five key messages.

- The first was **'Scotland's journey'** told through Scotland's rocks. Here large monoliths of rock represent all the key geological events in Scotland's geological and palaeogeographical history from deep in the Southern Hemisphere to its present position.

- A **'Slice through Scotland'** illustrates the terranes that make up Scotland's scenery with each bounded by major faults that are visible from space. These terranes characterize the differing styles of landform that make up Scotland's diverse landscape.

- **'Journey to the centre of the Earth'** is sited in a 'cave' in the centre of the amphitheatre and tells the story of the internal structure of the Earth; the crust, mantle and core and how convection in the mantle drives plate movements.

- The role of **James Hutton** and the connections between *Our Dynamic Earth* and the ancient volcano of Arthur's Seat has always been a part of the reason for *Our Dynamic Earth* being where it is. Display boards have been installed interpreting the geometry of the volcano and highlighting the place of James Hutton on the evolution of geological ideas.

- Finally, the **'Geogarden'** is a place to sit and contemplate these stories in an environment surrounded by plants with a geological significance.

The next chapter in the story of *Our Dynamic Earth* is what will the future be? This was the concept behind the development of the third bid to the ReDiscover fund. The FutureDome uses the full dome screen in what had been the Showdome in a new and exciting way. It lets visitors explore what the future of our planet might be, taking climate change as the overarching theme. There are some aspects over which humans have no control and the FutureDome takes you back in time to see the effects that huge volcanic eruptions and meteorite impacts have had on the planet in the past, as these will undoubtedly happen again in the future but we don't know when. But there are things that humans have a degree of control over. Future climate will be determined by how we use energy; available drinking water will be determined by future climate; our need for both these commodities will be determined by population numbers. Everything is inter-linked. Choices citizens make today will determine what the issues will be when we travel forward in time to 2030 and these will pose further choices which will raise further issues when we travel to 2055 and onward to 2080. The choices we make will determine possible futures. The FutureDome is a voyage of exploration, and is a world first combining new technologies of rotating floor pods and full dome projection where each visitor can express their own choices, to demonstrate interconnectivity of the different facets of the Earth's system including us and how our actions might affect the future.

Final thoughts

The documentation of the evolution of *Our Dynamic Earth* from its conception through to its state in 2008 highlights a journey from an idea to a reality with many of the high-lights and lowlights in between. It is an exercise in reflection and in sharing experience through nearly a decade of a major visitor attraction and science centre.

■ REFERENCES

Agassiz, L. (1849). A memoir to the author. In: *Footprints of the Creator* (Miller, H.) Nimmo, Hay and Mitchell, Edinburgh.

Bronowski, J. (1973). *The Ascent of Man*. BBC Publications, London.

Hutton, J. (1785). *Abstract of a Dissertation Read in the Royal Society of Edinburgh upon the Seventh of March, and Fourth of April MDCCLXXXV, Concerning the System of the Earth, its Duration and Stability*. Edinburgh.

Hutton, J. (1788). Theory of the Earth: or an investigation of the laws observable in the composition, dissolution, and restoration of land upon the globe. *Transactions of the Royal Society of Edinburgh*, 1, 209–304.

Meisner, R. and Osborne, J. (2009). Engaging with interactive science exhibits: a study of children's activity and the value of experience for communicating science. In: *Investigating Science Communication in the Information Age: Implications for Public Engagement and Popular Media* (ed. R. Holliman, E. Whitelegg, E. Scanlon, S. Smidt and J. Thomas). Oxford University Press, Oxford.

Miller, H. (1849). *Footprints of the Creator*. Nimmo, Hay and Mitchell, Edinburgh.

Miller, H. (1857). *The Testimony of the Rocks; or, Geology in its Bearings on the Two Theologies, Natural and Revealed*. Constable, Edinburgh.

■ FURTHER READING

- HM Inspectorate of Education (2007). *Review of the Contribution of the Scottish Science Centres Network to Formal and Informal Science Education*. HMIe, Livingston. Available at: http://www.hmie.gov.uk/documents/publication/sscn.pdf. A report conducted by the education inspectorate to evaluate and to give advice to the Scottish Science Centres Network on improving the learning, both formal and informal, within the centres.

- Monro, S.K. and Crosbie, A.J. (1999). The Dynamic Earth project and the next millennium. In: *James Hutton—Present and Future*, Geological Society Special Publication 150 (ed. G.Y. Craig and J.H. Hull), pp. 157–167. A paper giving an overview of the aspirations for the *Dynamic Earth* project in its early stages of development.

- Morrison, I. (1998). Makkin Siccar—Dr James Hutton on the rocky road to evolution. *COSMOS (Journal of the Traditional Cosmology Society)*, **13**, 69–89. A paper giving insight into the philosophical contribution of James Hutton to scientific thinking during the Scottish Enlightenment.

- Scottish Executive (2005). *A Four-year Strategy Developed by the Four Scottish Science Centres in Partnership with the Scottish Executive*. Scottish Executive, Edinburgh. Available at: **http://www.scotland.gov.uk/Publications/2005/12/06113103/31038**. This report defines the strategy for the development of the Scottish Science Centre Network over a 4-year period from 2005 to 2009.

◼ USEFUL WEB SITES

- *Our Dynamic Earth*: **http://www.dynamicearth.co.uk/**. The *Our Dynamic Earth* web site provides information on what is going on at the centre and summarizes the educational materials.

- Ecsite: **http://www.ecsite.net/new/**. This is a portal to science centres throughout Europe, also listing information on science communication conferences.

- STEMNET: **http://www.stemnet.org.uk/**. STEMNET works with a range of partner organizations, including government, industry, professional institutions, education and other major companies, to ensure recruitment of people with a range of science, technology, engineering and mathematics (STEM) skills.

- SETPOINTs: **http://www.setpointscotland.org.uk/**. This web site gives information on what is being done across the four SetPoints in Scotland. SetPoints are hosted by a diverse range of specially selected organizations skilled in facilitating links between education and the wider STEM community and working with other STEM partners. Operating at a local level, they provide access to activities and schemes for students and schools.

6.2

Engaging through dialogue: international experiences of Café Scientifique

Ann Grand

Introduction

The UK-based web site of the Café Scientifique movement defines a café as 'a place where, for the price of a cup of coffee or a glass of wine, people meet to discuss the latest ideas in science and technology which are changing our lives'.[1] This broad description and the flexible structure has allowed some 200 or so organizers of 'Cafés Scientifiques', 'Science Cafés', 'Bars des Sciences', 'Science Ka Adda', 'Wissenschafts-café' (to mention just a few of the designations of cafés world-wide) to adapt a simple model for public engagement with science to their local circumstances and culture.

In this chapter I provide a practitioner's perspective, based on 3 years' experience of organizing a Café Scientifique in Bristol, 3 years working with the English network of (junior) Cafés Scientifiques and the international network of (senior) Cafés Scientifiques and drawing on evidence from a recent international conference of Café Scientifique organizers. I will document some of the history of this relatively recent method for public engagement with science and describe various models of Cafés Scientifiques in different circumstances and cultures. In the latter sections I briefly discuss some of the rationale and challenges faced by those who wish to extend the Café Scientifique format into virtual territories, finally considering future developments for this still-evolving format.

Café Scientifique is an organization with very little organization; a loose network of independent people doing it because they are passionate about opportunities to engage with science, bound together by a set of common, core principles and boundless enthusiasm. There is no official registration or qualification, except a willingness to have

1. http://www.cafescientifique.org/.

a go. For this reason there are almost as many models of Café Scientifique as there are cafés but what is the typical experience?

Cafés Scientifiques take place in all sorts of venues—cafés, bars, pubs, theatres, restaurants, museums, arts centres, galleries—but the uniting feature is that wherever they happen, it is always outside a formal academic context. There is no sense or flavour of the lecture hall or laboratory demonstration; the venues are places where anyone can walk in off the street and feel comfortable in doing so. Entrance is usually free, with no pre-booked tickets; participants just turn up on the night. Most are funded simply by passing round the hat for voluntary contributions towards the speaker's expenses. Cafés Scientifiques classically start with a short introductory talk by an 'expert'. On average, this introduction lasts for no more than 25 minutes, and in some cases as few as 10. Most cafés eschew technology—so no slides, microphones or dimmed lights to emphasize a hierarchical divide between speaker and audience. Speakers are typically practising scientists, scholars or writers, actively involved in the field under discussion. Ideally, speakers should be drawn from as wide a range and area as possible; not just academia and medicine but also local businesses and industries.

After the introductory talk, there is a break, to allow glasses to be recharged and conversations to start. After the break comes an hour or so of discussion, questions, comments, thoughts and opinions between the speaker and the audience, the audience and the speaker and the audience and the audience. At its best, this interaction is spontaneous, dynamic, informal and as symmetrical as possible: much more 'dinner party' than 'after-dinner speech'.

How did it all start?

In the UK, the first Café Scientifique was held in Leeds, in 1998. It was started by Duncan Dallas, then an experienced television producer of science and medical programmes. Duncan had become increasingly irritated by what he saw as television's inability, or unwillingness, to tackle challenging contemporary ideas in and around science. In 1998, reading the obituary of Marc Sautet, founder of the French 'Café Philosophique' movement, it occurred to him that although Cafés Philosophiques had not taken off in the UK, Science—a capital-lettered subject that the British apparently took seriously—might be the basis for some interest. He persuaded the manager of a nearby bar to host an event, recruited a local scientist to be the speaker, put a poster in the window, told some friends and hoped for the best.

The meeting was considered a modest success at the time; some 30 people came along. Duncan organized another, also well-attended and was pleasantly surprised that participants were eager to know when the next events were to be:

... it slowly dawned on me that I had accidentally discovered a real appetite for science and discussion—not science as taught in schools but science as experienced and debated. There was also a real appetite among scientists to explain their work and they too seemed to enjoy these evenings.

Dallas (2006, p. 2)

What Duncan didn't know then was that cafés already had an international dimension, having started in France some 6 months before: this format was clearly timely. The first café in France was in Lyon, in October 1997 organized by a group led by Pablo Jensen, then a physicist at the Centre National de la Recherché Scientifique (CNRS) in Lyon. This group's initial concern (and indeed, the topic of their first café) was with methods of promoting public understanding of science:

... the sciences evolve and do not wait, most of the time, for non-scientists to understand them ... what is the goal of popularisation; its disadvantages, its effects? ... should the questions be asked by non-scientists or put forward by researchers?[2]

Although the group realized that 'two short hours' were not long enough to comprehend the whole of a complex issue:

... 'cafés sciences et citoyens' were effective means of exchanging ideas, establishing links between science and society and promoting better understanding of the world of science and research.[3]

From both these cities, cafés spread: from Lyon, across France and to a lesser extent to other countries in Europe; from Leeds, first to Nottingham and Newcastle and then across the UK. This early growth was slow, informal and gradual, transmitted by word-of-mouth rather than through formal, funded development: a speaker at one café took the idea back to his or her home town; one bar owner passed on this idea for filling up quiet nights to another.

When these first cafés started in the late 1990s, there were few formal venues or organized formats for public discussion of science. Then, in 2000, a House of Lord's report argued that: 'Society's relationship with science is at a critical phase . . . public unease, mistrust and occasional outright hostility are breeding a climate of deep anxiety among scientists' (House of Lords 2000, Introduction, para. 1.1). In the UK, the immediate causes of that mistrust were the BSE scandal and its links to vCJD (see Irwin 1995, 2009), Dolly the sheep and associated ideas about human reproductive cloning (see Holliman 2004) and concerns about the commercialization of genetically modified crops (see Thomas, Chapter 4.2 this volume); since then have come the public debates about nanotechnology, stem cell research, the MMR vaccination and more.

Since 2000, public engagement with science has become fashionable *and* fundable and 'informal dialogic' approaches (Davies 2009), such as Café Scientifique, have received support. Money has flowed from government, public bodies, research institutes, medical charities, media organizations and scientific societies to encourage innovation in developing methods of deliberation, dialogue, participation, horizon-scanning and outreach. The public engagement activities that have emerged from this funding are diverse: festivals, lectures, school-based activities, deliberative meeting of citizens (Democs),[4] citizen juries, online consultations and the UK-wide *GM Nation?* debate (to name but a few). Many of these activities require large-scale resources for their development and sustenance.

2. http://perso.ens-lyon.fr/pablo.jensen/.

3. http://www.1001-sciences.org/. Translated by the author.

4. http://www.neweconomics.org/gen/democs.aspx.

In contrast, Café Scientifique (given plenty of 'people-power' and enthusiasm from its organizers) is a self-sustaining, cheap and simple format. Although it has received some relatively small-scale funding (see later), each café is self-supporting and most operate on the thinnest of shoe-strings.

Oliver Sacks, the writer and neurologist, said the point of cafés was 'to bring science back into culture' (D. Dallas, personal communication). Some people see them as increasing public understanding of science, some as science communication, science engagement or even science education activities. The flexibility in the format allows for multiple aims and objectives to be sustained. However, because the liveliest, and longest part of a café is the discussion, Dallas believes cafés are really a:

> . . . cultural examination of science. The audience is . . . examining [science] from the outside, not the inside. This opens up a range of questions and issues which would not be covered by a debate within science itself.
>
> Dallas (2006, p. 3)

Wherever they are and whatever culture they operate within, the common principles of cafés are that:

> First, they move discussion into the public arena—academics go to the public, not the public to academics. Second, we [café organisers] have no brief to defend science at all costs. This provides a free and open agenda under which people are welcome to ask awkward questions. Third, there is face-to-face contact with scientists at a community level and finally, we are a network, not an organisation; bottom-up, not top-down.
>
> Grand (2007, p. 4)

As the Café Scientifique web site says, cafés are 'a forum for debating science issues, not a shop window for science'; they have no aim always to be supportive of science. Audiences make up their own minds, which is believed to suit the mind-set of the 21st century, in which people who want to take part in these types of events develop and contribute their own views, rather than have wisdom handed down from on high.

The common philosophy of cafés is that science impinges on every culture, world-wide, therefore citizens should have the right to engage with these sorts of issues. The science is pretty much the same but the cultural responses are different, hence the different formats, topics and locations. The over-riding value is for culture to engage with the scientific ideas and draw science back into conversation. The unique selling point of Café Scientifique is that changing the venue changes the nature of the debate to make it relevant to those who are participating: in a classroom, you expect to be taught, in a lecture theatre, you expect to get lectured at, but in a café, you expect to have a conversation.

Evolution

The organic, autonomous and independent nature of the Café Scientifique network means that it is always difficult to say exactly how many are active at any one time. At the time of writing (2008) there are approximately 40 cafés in the UK, 50 in North America, nine in Central and South America, 50 across continental Europe, 40 in Japan, six in other countries in Asia and the Pacific and one each in Uganda and Iran.

Some cafés begin independently, unaware of the existence of an international network and only find out about it later. Some cafés seem permanent fixtures, handed on from one organizer to another; others flourish then fade as their organizers move away. Survival often depends on practical factors; venues close, new management ends the association with the café or very small contributions mean it is impossible to pay speakers' expenses or create publicity. Having a dedicated group of organizers, rather than a single individual, has proved vital to those cafés that have endured: this group can share all aspects of organizing, and rotate roles. Written guidelines about every aspect of the café also makes organization smoother: from how any equipment works, to how to recruit, brief and introduce speakers and moderate the speaker/audience interaction.

In the UK

The early spread of cafés across the UK was organic and personal. Visitors to the Leeds café took the idea back to their home town; one speaker passed on the idea to another; enterprising bar owners saw opportunities. The second UK café was in Nottingham, started by a bar owner keen to extend the community involvement of her business; the third was in Newcastle, started by the sociologist Tom Shakespeare when he joined the university's Policy, Ethics and Life Sciences Institute.[5] The idea spread across Europe and to the USA, South America, Japan, Australia and New Zealand and, most recently, to tender new shoots in Pakistan, Iran and Uganda.

Café Scientifique's earliest major financial support came from the Wellcome Trust.[6] In 2002, under its 'Science and Society' public engagement programme, Duncan Dallas and Tom Shakespeare (by then both keen supporters of this format for public engagement) received a 3-year grant to extend the network. This grant funded two part-time coordinators who supported the development of new cafés across the UK.

The coordinators worked in two ways: one was to target specific towns where instinct or personal knowledge led them to feel cafés would be successful. They found venues and speakers and ran the café themselves until local organizers came forward. Although this led to cafés having a sound start, it was time-consuming for the coordinators and, given the face-to-face nature of the format, was only really practical in towns close to their homes. It was also crucially dependent on generating a willingness among participants to take over the running of the café. Alternatively, they supported individuals who approached them directly, having found out about Café Scientifique through the web site, word-of-mouth or media reports. Where feasible, the coordinators visited the potential organizers, supported them in getting their café started, often attended the first event and suggested possible speakers and topics. If this wasn't possible, they supported the organizers by phone and e-mail. New cafés received 'seed money' (typically less than £100) for costs such as publicity and speakers' expenses. The coordinators were also the link between the company that maintained the Café Scientifique web site and the organizers.

5. http://www.ncl.ac.uk/peals/.
6. http://www.wellcome.ac.uk/.

The development of the Café Scientifique web site was crucial in the growth of the network.[7] It became a point of contact, source of help and advice, and a public presence for the emerging network of UK cafés. Café Scientifique's presence on the internet also helped the idea spread world-wide; of their own volition, cafés started in, for example, Brazil, Costa Rica, Japan and the USA. Because individual Cafés Scientifique are run by volunteers, with no hierarchy or central organization, the web site remains vital in holding the network together. However, sustaining such a network is a concern of café organizers; at their 2007 conference, this was one of the topics discussed:

There is value in being part of a network: it adds legitimacy and validity to the café movement, it's comforting and exciting to feel part of a world-wide family, it's easier for new cafés to feel they have somewhere and someone to go to for help and it's possible to share resources, such as lists of speakers.

Grand (2007, p. 19)

Beyond the UK

Beyond the UK, the development of the Café Scientifique movement has very much depended on local cultural conditions. In Japan, for example, the number of cafés has grown very rapidly since the first was held in Kyoto in 2004; there are now between 40 and 50 regular cafés and special cafés are frequently held at conferences or during public events such as science festivals. However, their development has followed a subtly different pattern to that in the UK. Masaki Nakamura, co-organizer of a Tokyo café, described it at the 2007 international Café Scientifique organizer's conference:

In 2004, the government published a White Paper which introduced the idea of science cafés to the wider Japanese public. In 2006, the Science Council of Japan organised twenty science cafés in Science Week, which got science cafés more widely known. At that time, as a member of the Institute [The National Institute of Science and technology, part of the Ministry of Education, Culture, Sports and Science], I helped promote the model of science cafés as discussion and dialogue-centred events. Although top-down, these initiatives promoted the idea of science cafés rapidly in Japan.

Now, a wide variety of groups and organisations runs science cafés. Some are volunteers, like my group; some are individuals, fascinated by the café philosophy and some are run in bookshops, by their owners. But in a two-sidedness that represents the Japanese science café movement well, some are run by local and national government, as part of promoting public trust in science and technology. Some are run by academic societies, like the Science Council of Japan and by universities and research institutes as part of their outreach work. To facilitate grass-roots activities, we must think about the relationship with local and national government.

Quoted in Grand (2007, p. 6)

In the USA, the early development of cafés was very much like that in the UK—people heard about the idea, picked it up and made it work in their communities. While some cafés are run by interested individuals, there is also a degree of institutional support from national science organizations, which see Café Scientifique as a way of fulfilling *their* aims

---

7. http://www.cafescientifique.org/.

for public engagement, a development which may eventually conflict with the founding aims of Café Scientifique. Since 2005, the television channel WGBH, which produces *Nova ScienceNow*,[8] has employed a staff member specifically to promote the development of cafés, using a model based on *Nova* science programming (see later) and the programme hosts a new web site for American cafés.[9] Furthermore. SigmaXi, a scientific research society, encourages its members to start cafés and is working on having them as part of its 'distinguished lecture' series. At the time of writing (2008), the American Chemistry Society and the American Physical Society were considering adopting a similar approach. NASA encourages its scientists to present at cafés and works directly with organizers, and the National Science Foundation has a café at its Washington, DC headquarters and is thinking about specifying cafés as an outreach activity for its grant recipients. Taken together, these approaches suggest the emergence of institutional support for Café Scientifique. While this may secure the future of the movement in some countries and make it easier to secure speakers for café events, it may also transform the nature and purpose of the original format, potentially giving the perception that cafés are run by and for scientific institutions, rather than by and for the audience.

Another factor in spreading the café idea across the world was the British Council. Its science office (based in Manchester, UK), picked up on the concept very soon after its inception and promoted it through their network of science officers in British Council offices world-wide.[10] Like any other café organizer, the British Council has adapted the way it runs cafés to suit local circumstances: some offices host cafés run in English, using scientists and writers whom the British Council has invited to speak in the country; some host cafés in the local language with both native and non-native scientists; some host cafés using video-conferencing with the speaker in one country (for example the UK) and the host and audience in another. (This model is explored further later in this chapter.)

Junior cafés

As shown by the recent changes to the UK National Curriculum, setting science (and the issues arising from that science) in its social context is seen as increasingly important, especially for citizens coming to adulthood in the 21st century. Young people enjoy talking about ideas, so moving the participatory, interactive, contextualized Café Scientifique model into schools was a development well worth trying.

Like 'senior' cafés, *Cafés Juniors* started in Lyon, in France; they were coordinated by the *régions*. The first region was Rhone-Alpes, where the organization was run from the Centre for Culture, Science, Technology and Industry (CCSTI), at the Pole University of Lyon. Cafés mostly ran in *lycées* (with students between 15 and 18 years old) but some were held outside schools. The first experimental cafés began in 1999; by 2003, there were approximately 65.

8. http://www.pbs.org/wgbh/nova/sciencenow/.
9. http://www.sciencecafes.org/.
10. http://www.britishcouncil.org/science-cafesci.htm.

Members of CCSTI worked with a group of 10 or so students in the school to develop the cafés, including training students in skills such as chairing and organizing the meeting. (The cost of this training, plus costs of hiring microphones, etc, amounted to approximately €200 (about £120) per café. Although the schools made a modest contribution, most of these costs was met from government grants to CCSTI. With their help, the students decided on a topic and defined the profiles of three 'ideal' speakers (*cafés juniors* follow the French model—see below—of having a panel of speakers). CCSTI then found speakers who matched the profile as closely as possible. The students met the speakers beforehand and two of the students chaired the meeting. Cafés typically lasted for around 2 hours and took place in school time. Teachers had very little to do with the café; they attended but largely as observers.

From their base in Rhone-Alpes, *Cafés Juniors* spread across France, following national publicity in a speech made by Claudie Haigneré (Minister for Research and New Technologies and formerly an astronaut) at the launch of the 2002 Lyon Science Festival. Mme Haigneré commented:

. . . cafés in schools are becoming places to talk about big scientific problems . . . the informal and friendly dialogue with researchers brings the world of research closer to young people . . . to me, a pre-condition for the development of well-rounded citizens.[11]

Cafés Juniors are organized and funded by the *régions*; there is no national organization. This has led to problems of sustainability, and some areas have been more successful than others in keeping them going. For example, there are no longer any cafés in Rhone-Alpes. Once government funding stopped, the cafés stopped. However, in Île de France, *Cafés Juniors* continue to thrive, with some 30 organized by schools and a further 40 or so supported by the adult *Bars des Sciences* network.[12]

Cafés Juniors crossed to the UK in 2003 after Duncan Dallas attended the European Café Scientifique conference in Paris. Duncan organized a pilot project in Leeds. Five schools were approached and cafés were organized in three. Access to speakers was made easier by the fact that the local university's press officer was a regular at the 'senior' café; her contacts resulted in a dozen or so offers—from doctoral students to the Vice-Chancellor—to be speakers.

Following the success of the pilot scheme, in 2005 Dallas and Shakespeare successfully applied for a second 3-year grant from the Wellcome Trust to develop a programme of cafés in schools. This project is focused in the north of England and the first cafés took place in the autumn of 2005 with the aim of having 75 cafés by the end of the project (in mid-2008); at the time of writing, there are approximately 60. The greatest concentration of cafés is in the areas where the project began—Manchester, Newcastle upon Tyne, Leeds and Durham but this year we are turning our attention to new regions, especially around York and in South Yorkshire.

We have found that schools largely self-refer to the project: that is, teachers or students pick up the idea from the web site or from colleagues or friends in other schools and contact us directly. There has been very little media reporting of the project. This has led

11. http://www.1001-sciences.org/cafe_juniors/reseau.htm.
12. http://bardessciences.net/.

to the majority of the cafés being in thriving schools in leafy suburbs, therefore mainly involving children from middle-class and privileged backgrounds. This is an issue that we constantly try to address. Although we have no specific agenda to bring science to deprived areas and under-privileged students, we are concerned to spread the benefits of Café Scientifique. To this end, we have worked with organizations such as AimHigher to target certain schools, for example in Leeds.[13]

The founding principles of Café Scientifique in schools are, as you might expect, very similar to the original principles of the senior café network:

Café Scientifique offers a unique opportunity for students to meet working scientists in an informal, relaxed, 'café' atmosphere and together explore contemporary issues in science and technology. The cafés are student-led, from choosing the topic to running the café, giving those involved the chance to turn concerns into participation, based on classroom experiences. Cafés are open to all secondary schools: teachers and students, any age, any status, any interests . . . cafés take place in cafeterias, common rooms or libraries, not classrooms; at lunchtime or after school, so that audience and speaker meet as equals, without barriers.

Junior Café Scientifique web site (2005; http://www.juniorcafesci.org.uk/)

In line with recent changes to the National Curriculum (see Holliman and Thomas 2006 for discussion), the basic aim of the project is to help students, through informal conversation with working scientists, to develop their skills in thinking critically about and engaging with those issues in science and technology that will affect their (and our) futures. Though this is our main aim, teachers have reported beneficial side-effects, such as developing students' understanding of the science curriculum, better understanding of the philosophy and workings of science and contributing to their empowerment and entrepreneurial skills.

Like its 'senior' predecessor, the growth of Café Scientifique in schools is characteristic-ally bottom-up because it is participant-led. There has been no advertising (other than the web site) and very little media coverage; our growth has almost entirely been fuelled by word-of-mouth and electronic communication. Once a school has expressed an interest (typically, though not always, through a science teacher) the project organizer visits the school. She runs a 'taster' café (acting as speaker herself) for any interested students or those identified by the teacher as potential organizers and audience. This is followed up by a practical workshop to help the students set up an organizing team and start planning their café programme. After that, the café is largely independent and self-sustaining, although the project organizer's support remains considerable, chiefly in helping the student organizers find speakers that match the topics they have chosen.

Again like their 'senior' relatives, the culture of each school café is subtly different. Some run at lunchtime, others after school; venues vary and so do numbers—in some schools, just six or so people turn up for each café, which leads to an intense discussion; in some schools, 40 or more students form the audience; in a few cafés, refreshments are commercially sponsored (branded coffee and buns bring in the crowds) while others run on orange squash and biscuits provided by the school. One contrast to senior cafés is that schools equivalents are much shorter; typically lasting about 40 minutes.

13. http://www.aimhigher.ac.uk/home/.

Popular topics for discussion have included topical controversial life science issues, for example therapeutic and reproductive cloning, animal/human chimeras, genetic engineering and the teaching of creationism and evolution in schools, plus 'sci-fi' subjects like parallel universes, time travel, aliens, artificial intelligence and computer games. Less obvious subjects have also been the subject of successful cafés, for example the science of colour and deep sea life. The most successful cafés are those where the speaker finds a way of linking the topic to everyday teenage life.

Though France and the UK are the two countries with the most developed network of *Cafés Juniors* in school, there are signs that this format is spreading. There are a few junior cafés in the USA, some in Japan and one in Germany. Unlike in the UK, cafés in these countries are generally held outside school and school hours—typically on Saturday morning in a venue such as a museum café. This change of venue is a very significant difference, the impact of which is often overlooked. People sometimes suggest this approach to us, as it takes cafés into new environments and we have ourselves organized them at science festivals, conferences, etc. and it certainly leads to creative encounters. However, taking the café out of school removes the crucial element of student ownership, as any event outside school requires that teachers control planning, venue, transport and so on. The cafés are no longer open to 'passing trade' from within the school, just the 'usual suspects' and those well-organized students who remember to put their name down in time. In other words, the cafés cease to be bottom-up and participant-led and this primary virtue no longer applies.

Speakers, topics and audiences

One of the crucial issues to address when organizing a Café Scientifique is the choice of speaker. As the Café Scientifique conference report argues:

Who are the 'right' speakers? Should we choose speakers who will provoke the audience or do audiences prefer to be spectators? Do we want zealots or a more detached perspective? Large crowd-pullers or younger scientists, talking about their current work?

Grand (2007, p. 11)

All cafés want 'good' speakers but, oddly enough, the quality of the speaker is not really important. Although having a good speaker helps, a bad speaker doesn't necessarily ruin a café. The audience, rarely the speaker, makes the evening. If the subject is interesting or controversial and the audience is confident (or is led by a confident host) then the discussion is likely to be worthwhile. In many ways, the speaker is only there to give the audience enough information to start asking intelligent questions, to respond to audience comments and to help generate the conversation. (It's worth noting that although I talk about 'speaker' in the singular, in France, Denmark and elsewhere— for example, occasionally in the USA—cafés have multiple speakers or panels; thus the debate is less about the work of one expert and draws on different perspectives, meaning that the conversation tends to be less about the science and more about the context of that science within society.) Most speakers are practising scientists, talking about their current work. This is important; it maintains the unmediated relationship between the scientist (and the science) and the audience, means that the audience can ask questions about the science and be assured of a contemporary, authentic and assured answer and keeps the passion alive and the subject matter fresh and up-to-date. However, cafés also occasionally invite writers, journalists or scholars from other fields altogether, such as artists and poets.

Those who speak at cafés do so for all sorts of reasons. Some believe public engagement to be a responsibility and see cafés as a way to meet these goals. In an evaluation of the US café network carried out for *Nova ScienceNow*, 90 per cent of speakers said they were initially recruited by an organizer but 92% were interested in doing it again.[14] The survey results also noted a significant 'ripple' effect on speaker recruitment, as scientists speak favourably of their experience to their colleagues.

Given the informal nature of cafés, any advice given to speakers is very much down to the individual organizers. Experienced organizers (and experienced audiences) can ensure that speakers fit in with the original café ideals—however they interpret them. However, in supporting people who are thinking about starting new cafés, I have found it necessary to be a little more prescriptive: strongly discouraging the use of slides; encouraging a short and succinct introductory talk, leaving plenty of gaps where the audience can envisage themselves placing questions and developing the notion that cafés are about conversations *between* audiences and speakers, rather than a linear transmitter–message–receiver model of communication (as described by Leach *et al.* 2009).

This relaxed and autonomous attitude towards 'training' works well for speakers talking to adult audiences, but in the schools' cafés we have supported speakers much more directly. School audiences, by their nature, may be less willing to challenge speakers—students are more used to being taught didactically by adults, not necessarily to questioning the premises of those who stand in front of them. So, in schools' cafés, it is much more up to the speakers to generate the vital atmosphere of conversation and intelligent questioning. We support them in two ways: through a speaker pack, downloadable from the web site, with full information about the aims and methodology of Café Scientifique, and through speakers' orientation workshops that introduce speakers to the aims, objectives, style and format of Café Scientifique and help them develop communication skills for presenting effectively to school students.

Topics

All cafés offer their audiences the opportunity to influence the choice of topics, and speakers may themselves be members of the audience. Such an approach is in keeping with the bottom-up participant-led ethos of Café Scientifique. The general feeling of café organizers is that topics need to 'arouse curiosity . . . be slightly provocative . . . relevant to the public' (Balling and Schuler 2004, p. 46). Within those broad themes, there's plenty of scope for individuality. Notwithstanding the topics listed above under the *Cafés Juniors* section, a glance at the 'previous events' pages of the Café Scientifique web site will show

14. http://www.sciencecafes.org/cafe_impacts.html.

that there is barely a scientific topic that hasn't been addressed in a café somewhere, sometime. Much depends on the organizers' contacts and the resources available locally, but perennially popular topics include anything to do with neuroscience, consciousness and psychology, genetics, health and medical topics and environmental issues. What these topics have in common is that they are either personally applicable (consciousness is something on which we all feel we can offer an opinion) or currently relevant, topical and possibly controversial.

Audiences

Who goes to cafés? As the previous discussion about *Cafés Juniors* showed, there is anecdotal evidence to suggest an emphasis on 'leafy suburbs'. Such evidence may also be supported by formal surveys. The *Nova ScienceNow* survey found that the audience tends to be 'well-educated' with many attendees returning regularly. Forty-two per cent are aged 18–34, the sex split is approximately 50/50, most audience members are White and well-educated and 59 per cent are working or studying in a science-related field. Typically, in the Lyon café, a third are regulars, a third scientists and a third casual droppers-in (Grand 2007, p.9). In an informal survey of international café organizers before our recent conference (carried out in February 2007) I found attendances varied from 10–20 to 70+ per café; average attendance was 46–56.

Café formats

The UK model

The UK model format for cafés arose accidentally and, as it happens, fortuitously. At the very first café Duncan Dallas organized in Leeds:

. . . as the speaker started to talk my heart sank to my boots: he talked in academic jargon as though addressing students. Fortunately, after his twenty-minute talk we had to have a break for drinks, as the coffee machine had been turned off during the talk because it was too noisy. After this break and once the audience started to ask questions, the evening improved rapidly. The speaker began to realise who he was talking to and what he had to explain. The questions were simple, direct and often difficult and the speaker began to think on his feet, always an enjoyable spectacle for the audience.

Dallas (2006, p. 1)

Thus came about the simple and robust short talk–break–discussion format, with a single speaker, which is still the pattern followed by most UK and Anglophone cafés. The short talk introduces the topic; the break allows participants to start conversations and think about issues to raise and the discussion gives everyone who wants to the chance to join in with a thought/opinion/comment/question. Having just one speaker means that the conversation with the audience is maintained; if there is a panel of speakers, the audience can become auditors or viewers of, rather than participants in, the debate

(Davies 2009). On the other hand, it does mean speakers have to be prepared to deal with criticism or opposition and to field questions from all points of the topic's compass. They do, however, always have the option of saying, 'I don't know'.

Most cafés have some kind of 'host' (facilitator, chair, coordinator, moderator—call them what you will). The level of control the host exerts depends very much on their interpretation of the role (Davies 2009) but they are very important to the discussion section—keeping conversation circulating, making sure everyone has the chance to contribute, keeping the dominators subdued, being prepared to jump in if conversation lags and making sure the speaker doesn't turn the discussion session into a chance to extend their talk. This is not universal—some cafés (for example, San Francisco's 'Ask a Scientist' café) deliberately extend democracy by not having a host at all. This works very well in intimate venues with smaller audiences but does place a tremendous responsibility on the speaker to keep things moving.

The success of this model is very dependent on the audience's willingness to join in discussion, debate and argument, and there are undoubtedly cultures where this is not always present. For example, the British model largely prevails in Japan, where people are not so well-used to a culture of discussion. In Japan:

> . . . organisers have to think about ways to facilitate this. Many organisers feel that an audience of no more than twenty to thirty is desirable. Some have experimented with breaking the audience up into groups of five or six people, as people may feel more comfortable talking in smaller groups. And some take questions by text or email. Some have experimented with 'voting' by mobile phone and some have connected several venues over the Internet.
>
> Grand (2007, p. 6)

In the USA, although cafés largely follow the UK model, because the television station WGBH encourages the development of science cafés, some include short video clips. Typically less than five minutes long, shown at the start of the event, these set the background for the speaker and the topic.[15]

The Danish model

Danish cafés emphasize the 'cultural and societal', promoting 'interdisciplinary discussions across the natural and social sciences, humanities, art and culture' (Balling and Schuler 2004, p. 24) The cafés involve two speakers, often from opposing sides of a question or from widely differing fields. For example, a café on the topic of 'The clones are coming' had as speakers a philosopher and a scientist so that the different perspectives were juxtaposed. Having two speakers is an attempt to:

> . . . expose the audience to experts who can put forward a diversity of perspectives in an enthusiastic manner and who can have an appreciation for the other experts on the panel—often, panellists have limited knowledge of each other's expertise. We believe that this method creates ideal conditions for a reflective and nuanced dialogue in an open-minded environment.
>
> Balling and Schuler (2004, p. 29)

15. http://www.sciencecafes.org/.

For this to be successful, the panellists must establish a good relationship and become reasonably well acquainted with each other's fields of research. To this end, they meet, together with the moderator for dinner before the café to start this conversation.

Danish cafés last around 90 minutes, without a break. After a short introduction, the panellists have eight minutes to introduce themselves personally and professionally. The remaining 70 or so minutes are devoted to questions, which may come from either the panel or the audience. The role of the moderator is crucial; in the Danish model they must control 'shop-talk' and jargon and ensure that the audience becomes involved from the very beginning, whether by commenting, asking questions or answering them.

This model has received many plaudits, including nomination for the EU Science Mediation award in 2007 and the Svend Bergsøe Award for Innovative Science Mediation in April 2007.

The French model

Like the Danish model, French cafés involve more than one speaker—often as many as four or five—who sit on a panel drawn from a range of organizations. French organizers value alternative accounts when presenting information, so the panel might include a scientist, a businessperson, a politician and an activist:

Inspired by the scientific practice, the Science Café's organizers wish to hear ideas from all parties, giving the opportunity to debunk preconceived ideas. Of course, technical experts are always part of debates . . . scientists should not be left by themselves when it comes to discussing the relevance of their scientific arguments in the public space . . . activist groups such as Greenpeace—possibly with their own experts—or healthy volunteers used for drugs testing for example are welcome to participate in Science Café discussions.

Balling and Schuler (2004, p. 35)

There is no introduction; each panellist has two minutes to detail their position in relation to the evening's topic, after which the audience is invited to ask questions. Cafés last one-and-a-half to two hours. All the participants—panellists and audience—sit at café tables, to encourage decisions and processes that are as democratic as possible.

Cafés and digital technologies

Cultural differences are not confined to national borders. Technology is also influencing both public engagement in general (Holliman 2008) and the Café Scientifique format. The relationship between cafés and digital technologies has often been uneasy, however. Given that many organizers eschew the use of old technologies, such as presentation software and microphones, video-conferencing, webcasts, podcasts and blogs might just be a step too far.

There is rationale behind this uneasiness: the essence of Café Scientifique is in the informal, relaxed, conversational and above all direct and immediate relationship between speaker and audience members. Many organizers feel that these technologies impose implicit and explicit barriers between participants, making it much harder for the audience and speaker to absorb non-verbal communication. Video-conferenced cafés, for example, face this problem of lack of personal contact between audience members and speaker. Speakers find it hard to get a sense of what the audience is like and the audience doesn't get the full experience of 'reading' the speaker—body language, for example, is curtailed. Having cameras on the audience also means there is a lack of intimacy and possibly as a result less willingness to ask candid questions. Furthermore, there are issues of privacy and copyright to be considered whenever a recording is made.

However, video-conferencing and webcasts are undoubtedly ways in which cafés can reach out to a wider more geographically distributed audience, who perhaps live beyond the easy reach of an actual bar or café or are constrained by location or circumstance. For example, the organizer of the Danish café, Gert Balling, has experimented with using CERN's 'grid' computer technology to create multi-national, multi-lingual, interactive virtual cafés.[16]

Cafés can also encourage intercultural dialogue and give people the opportunity for collective debate. The British Council has used such technologies widely, hosting cafés where the speaker is in one country and the audience and host in another, for example in Palestine, a country where there are few actual cafés, little freedom of movement and an ever-present threat of violence. Contrary to the usual non-academic venues, these cafés are held at a local university (a relatively safe place) with video links to the UK connecting the participants to the speaker. After the introductory talk, local moderators start a discussion from which questions are put back to the speaker. Thus, interaction and discussion are encouraged in a difficult environment.

How much or how little technology is used lies, in the end, with the individual café, with the enthusiasms and abilities of the organizer(s) and, to a lesser extent, the audience and speaker(s). Such is the strength of the robust and flexible Café Scientifique format. For example, although it no longer exists, the San Diego (California) café used to webcast its cafés because the organizer was involved in a webcasting company. The Macafé Scientifique (Hamilton, Canada) café has podcasts of some of its cafés (however, there is an issue of how much or how little to edit recordings of these events to provide listeners with an authentic experience),[17] the Southern California café has a blog[18] and Louisville (Kentucky) has a Facebook account.[19]

16. Dialogue in Cyberspace, available from http://www.caféscientifique.org/inthemedia.htm.

17. Available from http://www.slackerastronomy.org/wordpress/index.php/category/audio-podcasts/.

18. http://www.socal-sciencecafe.org/blog/.

19. http://www.cafescientifique.org/louisville.htm.

Future developments

The last 10 years have shown the format of Café Scientifique to be robust, adaptable and flexible, allowing for many cultural, local and online variations. A session at the recent organizers' conference (Grand 2007) revealed a wide variety of experiments on the 'straight' café format: pairing speakers, such as a scientist and an artist, quizzes, science games, experiments, science cabarets, play-readings of science-themed plays and a 'theory slam'.[20] Additions to the basic café were such things as encouraging speakers to give simple, 'string and jam-jar' demonstrations, fancy-dress, themed drink and food, laboratory visits and trips to the cinema to watch a science film, followed by discussion. Such innovations extend the traditional format of Café Scientifique, sometimes merging with other existing formats, and have the potential to attract new or more diverse audiences, ensuring that organizers have new challenges to engage with, hopefully adding value to the communal experience.

Continued independence versus formal sponsorship is a perennial issue for a network that survives on minimal funding. Most cafés are happy to operate on a hand-to-mouth basis and self-funding can be a great advantage for those with sufficient energy, enabling café organizers to work to their own 'local' timetables and agenda, which may be very different from those of possible sponsors. Beyond the individual café, the growth of the UK café network (and its ripple effect on the world network) was supported by the Wellcome Trust grant—and the second grant for the development of junior cafés has meant (as a side-effect) that we have been able to maintain a central coordinator and web site, to facilitate communication between cafés world-wide. But this funding is not unending, and if the café network is to continue stably or to grow, funding, or sponsorship, will need to be secured. There are many different kinds of sponsorship: community, public, academic and institutional sponsorship would be likely to leave the ethos of the network unaffected but corporate sponsorship might be more problematic.

As Duncan Dallas has argued:

Café Scientifique has no self-serving agenda. It has obvious aims—moving science into the public arena, face-to-face interaction between scientists and ordinary people, the freedom to ask awkward questions, etc. If cafés are a cultural examination of science, then they are a wider cultural force, not a narrow political or economic one. Cafés are local, voluntary communities of interest and how they encourage cultural change will depend on the café itself. There is no grand plan or final outcome. What happens depends on the initiatives individual cafés are willing to take.

Quoted in Grand (2007, p. 13)

Café Scientifique is a small-scale international success story. Its simple, robust and flexible format means it can find a home across countries and across cultures. Cafés are run by and for the audience; scientists (and, on the whole, their institutions) are willing contributors, rather than leaders. When organizers move on or the audience stops coming, the café ends but the network absorbs such changes and continues to evolve.

---

20. Analogous to the more familiar 'poetry slam', participants have a short time—2 or 3 minutes—to propound a favourite theory, with audience votes deciding the best one of the evening.

■ REFERENCES

Balling, G. and Schuler, E. (2004). *The Science Café: Art, Culture, Science*. Hovedland Publishers, Højbjerg, Denmark.

Dallas, D. (2006). *Café Scientifique*. Paper presented at the AAAS Conference, St Louis, February 2006. Available at: **http://www.cafescientifique.org/downloads/aaas_talk.pdf**.

Davies, S. (2009). Learning to engage; engaging to learn: the purposes of informal science–public dialogue. In: *Investigating Science Communication in the Information Age: Implications for Public Engagement and Popular Media* (ed. R. Holliman, E. Whitelegg, E. Scanlon, S. Smidt and J. Thomas). Oxford University Press, Oxford.

Grand, A. (ed.) (2007). *Cafe Scientifique Organisers' Conference 2007*. Available from: **http://www.cafescientifique.org/downloads/conference%20report.pdf**.

Holliman, R. (2004). Media coverage of cloning: a study of media content, production and reception. *Public Understanding of Science*, **13**, 107–30.

Holliman, R. (2008). Communicating science in the 'digital age': issues and prospects for public engagement. In: *Readings for Technical Communication* (ed. J. MacLennan), pp. 68–76. Oxford University Press, Toronto.

Holliman, R. and Thomas, J. (2006). Editorial. *Curriculum Journal*, **17**(3), 193–6.

House of Lords, Select Committee on Science and Technology (2000). *Science and Society*, Third Report. HMSO, London.

Irwin, A. (1995). *Citizen Science*. Routledge, London.

Irwin, A. (2009). Moving forwards or in circles? Science communication and scientific governance in an age of innovation. In: *Investigating Science Communication in the Information Age: Implications for Public Engagement and Popular Media* (ed. R. Holliman, E. Whitelegg, E. Scanlon, S. Smidt and J. Thomas). Oxford University Press, Oxford.

Leach, J., Yates, S. and Scanlon, E. (2009). Models of science communication. In: *Investigating Science Communication in the Information Age: Implications for Public Engagement and Popular Media* (ed. R. Holliman, E. Whitelegg, E. Scanlon, S. Smidt and J. Thomas). Oxford University Press, Oxford.

■ FURTHER READING

- Grand, A. (ed.) (2007). *Cafe Scientifique Organisers' Conference 2007*. Available from: **http://www.cafescientifique.org/downloads/conference%20report.pdf**. This is a report of the Second International Cafe Scientifique Organisers' Conference, held on 12 and 13 May 2007, at Thackray Medical Museum, Leeds, UK. The report includes reflections on junior and adult Café Scientifiques in Palestine, Japan, the USA and various countries in Europe.

■ USEFUL WEB SITES

- Café Scientifique homepage: **http://www.cafescientifique.org/**. This is the homepage of the Café Scientifique network. It is described in this chapter as 'a point of contact, source of help and advice, and a public presence for the emerging network of UK cafés. The homepage includes a link to the junior café web site: **http://www.juniorcafesci.org.uk/**.

- **Deliberative meeting of citizens (Democs):** http://www.neweconomics.org/gen/democs.aspx. Democs is described by its developers, the New Economics Foundation, as 'part card game, part policy-making tool that enables small groups of people to engage with complex public policy issues'. The materials required to conduct a Democs activity are freely available from this site, following registration.

- **Living Knowledge: the International Science Shop Network:** http://www.scienceshops.org/. This is the web site of Living Knowledge, the international science shop network. The site describes a science shop as an entity that: 'provides *independent, participatory research support in response to concerns experienced by civil society'*. The site contains a range of useful information, including reports and access to the Living Knowledge journal.

Final reflections . . .

Richard Holliman, Jeff Thomas, Sam Smidt, Eileen Scanlon and Elizabeth Whitelegg

The chapters in this volume demonstrate how influential the practices of science communication are to the shaping of modern science. What is clear is just how different these practices are, not just in relation to their objectives but also in terms of who has responsibility for communicating, and the various forms of professional, 'pro-am' (Leadbetter and Miller, 2004) and citizen expertise that they bring to these acts. Scientists, of course, continue to provide the essential under-pinning for much science communication, but as both Gregory and Doubleday (Chapters 1.1 and 1.2, respectively) illustrate, the norms and conventions that govern their communication practices may be changing to match the demands of 'mode 2' or 'post-academic' science (Gibbons *et al.* 1994; Ziman 2000). Practising scientists are now required to (re)learn a range of professional communication skills, not least in retrieving information electronically, also relying on other forms of expertise, including that provided by information professionals (Gartner, Chapter 3.2) and patent lawyers (Schulze, Chapter 1.3).

More generally, it has become a truism that scientists require access to scientific information in a range of forms to be part of what Gregory (Chapter 1.1) calls 'the scientific community'. But the usability of scientific digitally formatted information is far from equally distributed. As Montgomery notes (Chapter 3.1), a routine lack of access to scientific information—the so-called 'digital divide'—and the continued emphasis on 'English' as the *lingua franca* of science have profound consequences for those who are excluded, or who may find it harder to 'pass' peer review because they are not working for a 'recognized institution' or are writing in a second language (Wager, Chapter 4.1). Of course, these forms of exclusion may also be partly due to the radical nature of the ideas under consideration; the desire not to stand on the shoulders of giants, rather to introduce new ideas and new perspectives that may take time and considerable rhetorical powers of persuasion to be accepted. To push the metaphor a little further, some giants may be more willing than others to be superseded. In these instances, as Lewenstein (Chapter 5.1) notes, the concept of a *festschrift*, to be published towards the end of a researcher's career, may be especially significant; a form of *post hoc* 'self-correction' to a specialism's scientific corpus.

But access to verified scientific information is not the only issue to be considered. Given the 'worth'—potential or realized—of scientific intellectual property within a globalized knowledge-based economy, the communication of post-academic science is often about the strategic management of information; knowing when (and when not) to communicate has become at least as important to professional practitioners as knowing how to communicate, and who has relevant expertise in particular areas. If it ever were the case, the days when science communication was under the exclusive control of scientists is long gone. As Schulze shows (Chapter 1.3) the protection of intellectual property, for example through patents, introduces additional forms of expertise other than science to this particular form of science communication. Such forms of communication place new demands on at least some scientific researchers. In certain areas of science, notably the life sciences, it is no longer sufficient to search the relevant corpus of peer-reviewed literature—patent searches may also need to be conducted and affected scientists need to re-skill to meet these new requirements.

It follows that notions of openness and transparency need to be (re)examined in the light of the 'dialogic turn' (for an example of dialogue in action, see Grand, Chapter 6.2; for a theoretical rationale, see Sections 1 and 2 of Holliman *et al.* 2009) *and* the demands of post-academic science. Such norms are inevitably challenged by the requirements on occasion to strategically withhold information, or release 'partial' heavily mediated accounts, for particular reasons, to certain audiences, and at specific times. As a result, the academic paper (Wager, Chapter 4.1) will look different to a patent application (Schulze, Chapter 1.3), a blog post (Chalmers, Chapter 2.2), a radio broadcast (Redfern, Chapter 5.3), an exhibit in a science centre (Monro, Chapter 6.1), a popular science, or science-in-fiction book (Lewenstein, Chapter 5.1 and Turney, Chapter 5.2, respectively), and so on. The chapters in this book make inroads into questions such as 'Why is it that these forms of communication are as they are?'; our challenge to the reader is to explore these forms of science communication in more detail.

All this may seem a long way from the widely held view, at least the one apparently held by many contemporary practising scientists, that communicating scientific findings via peer-reviewed publications and conferences is the main vehicle that drives 'science'; or should that read 'innovation' in the context of post-academic science? Indeed, notwithstanding some of the reservations raised by Wager (Chapter 4.1) peer-reviewed research papers are often cited as the 'gold standard' of reliable and valid scientific knowledge. And yet, as Gregory noted (Chapter 1.1), as a proportion of the overall number of practising scientists, as last in terms of the figures provided by UNESCO in 2005, relatively few choose to submit their work for verification in this way. Industry-based scientists, to which we can add those working for the state, either as government scientists or for the military, etc., may 'publish' their work, but not necessarily in an open and transparent manner for the scientific community to begin the processes of 'falsification', as described by Popper over 50 years ago (see Chalmers 1999; Yearley 2005 for discussion).

Of course, there are also emerging opportunities for scientists to communicate via web 2.0-inspired technologies, both within—for example, through preprint servers, such as arXiv (Chalmers, Chapter 2.2)—and across disciplines (Schummer, Chapter 2.1). These more immediate forms of publication offer scientists opportunities to communicate largely unrestricted by spatial and temporal constraints—globally distributed networked communities of small numbers of specialist scientists can blog about their latest findings and challenges, collaboratively author publications, and so on. To some extent this has the potential to make the workings of science more visible; to begin to challenge the limitations of the scientific paper that Medawar (1999) so adeptly identified. And yet this can come at a price. Whilst blogs may be directed at small numbers of scientists working in the same specialist field, they are also forms of *public*ation, with the deliberate emphasis on the public. In areas of frontier science—in particular those to which media professionals may already be sensitized because they are potentially 'newsworthy' and in which scientists have a unerring habit of eventually resolving the problems they set themselves, meaning that a story is likely to break somewhere down the line—these forms of communication can become sources of science news (see Chalmers, Chapter 2.2 for an example; Redfern, Chapter 5.3, for a media professional's perspective; and Allan 2009 for a theoretically informed discussion of how science becomes news.)

As Allan (2009) and Thomas (Chapter 4.2) have shown, such examples are often subjected to public scrutiny via news media, in part because they are controversial, as opposed to consensual. The ongoing tension between controversy and consensus characterizes much of what we might call 'frontier science'. In these instances, and in particular in the ensuing communications, scientists often defer to the notion of scientific consensus. Of course, ideas about scientific consensus are more easily imagined and spoken of than they are formally codified; notwithstanding the ongoing work of the International Panel on Climate Change (IPCC), online petitions in support of the idea that the HIV does cause AIDS, such as the Durban Declaration, and various examples of open letters often campaigning on single issues.

Thomas' arguments reinforce the point that 'science-in-the-making' (Latour 1987)—that which is novel and therefore more open to interpretation and consequently also uncertain, incomplete and potentially contested—is often also seen to be controversial because it can challenge what is considered to be the 'consensus'. And these controversies can be both 'scientific'—in terms of the scientific claims being made—and 'science-based' —because of the social and ethical implications of the work (Brante 1993). It follows that studying the communication of the science behind controversies is well worth a detailed examination, but Thomas goes further, arguing that contemporary scientists who become embroiled in science-based controversies are increasingly taking on pro-active and persuasive roles in the emergence and closure of controversies, enacted through various and sometimes novel forms of communication. (This analysis proves an interesting contrast with the 'internal' and largely historic scientific controversies considered by Collins and Pinch (1998).)

Given their high-profile, sometimes deserved, public controversies are an important part of the 'culture of science'. As Lewenstein (Chapter 5.1), Turney (Chapter 5.2), Redfern (Chapter 5.3), Monro (Chapter 6.1) and Grand (Chapter 6.2) argue, in some instances explicitly, in others implicitly, science informs culture and culture informs science; they are inextricably enmeshed. Having 'spaces' for engaging with the culture of science, such as *Our Dynamic Earth* and Café Scientifique, or other forms of science communication, for example, popular science books, science-in-fiction and radio are an important manifestation of this culture. We argue that scientists clearly have a critical, often central, role in practising these and other forms of science communication, but other forms of knowledge and expertise can and should have their place in these ongoing discussions and deliberations.

■ REFERENCES

Allan, S. (2009). Making science newsworthy: exploring the conventions of science journalism. In: *Investigating Science Communication in the Information Age: Implications for Public Engagement and Popular Media* (ed. R. Holliman, E. Whitelegg, E. Scanlon, S. Smidt and J. Thomas). Oxford University Press, Oxford.

Brante, T. (1993). Reasons for studying scientific and science-based controversies. In: *Controversial Science; From Content to Contention* (ed. T. Brante, S. Fuller and W. Lynch). SUNY Press, Albany, NY.

Chalmers, A. (1999). *What is This Thing Called Science?*, 3rd edn. Open University Press, Milton Keynes.

Collins, H. and Pinch, T. (1998). *The Golem; What Everyone Should Know About Science*, 2nd edn. Cambridge University Press, Cambridge.

Gibbons, M., Limoges, C., Nowotny, H., Schwartzman, S., Scott, P. and Trow, M. (1994). *The New Production of Knowledge; the Dynamics of Science and Research in Contemporary Societies*. Sage, London.

Holliman, R., Whitelegg, E., Scanlon, E., Smidt, S. and Thomas, J. (eds) (2009). *Investigating Science Communication in the Information Age: Implications for Public Engagement and Popular Media*. Oxford University Press, Oxford.

Latour, B. (1987). *Science in Action*. Open University Press, Milton Keynes.

Leadbetter, C. and Miller, P. (2004). *The Pro-am Revolution: How Enthusiasts are Changing Our Economy and Society*. Demos, London.

Medawar, P. (1999). Is the scientific paper a fraud? In: *Communicating Science: Professional Contexts* (ed. E. Scanlon, R. Hill and K. Junker). Routledge, London.

Yearley, S. (2005). *Making Sense of Science*. Sage, London.

Ziman, J. (2000). *Real Science*. Cambridge University Press, Cambridge.

■ INDEX

Note: 'n.' after a page number indicates the number of a note on that page.

A

abstracting services 99
Académie des Sciences, Paris 6
Académie Royale de Médicine 118
Agassiz, Louis 198
AimHigher 217
Alicki, Robert 72
AlphaGalileo 188
Amazon 106
American Association for the
    Advancement of Science
    (AAAS) 87
American Chemical Society 63
American Chemistry Society 215
American Physical Society 215
Anderson, Philip 77
Annual Reviews 152
Arcadia Biosciences 38
archives, e-print 90, 104–5
    see also arXiv
Armstrong, Lyle 137, 142
arXiv 12, 68, 69, 75, 79, 90,
    103–4
    blogs 72, 73
    peer review 126
    web site 111
Asimov, Isaac 159
Association of Learned and
    Professional Society
    Publishers (ALPSP) 119
asteroid impacts 185
ATLAS 77
Atmospheric Chemistry and Physics
    71
Attenborough, Sir David 160
Atwood, Margaret 169, 170
audio *see* podcasts; radio and
    other audio
audio-visual resources, online 92
Australia, patents 44
Aymar, Robert 74

B

Bacon, Dave 72
Balling, Gert 223
Banks, Michael 70

Banville, John 172
Baxter, James Phinney 156
Bayh–Dole Act 1980 (USA) 43,
    44, 46
BBC radio programmes 178
    historical context 179–80
    information age 188, 189, 190
    listeners 181–3
    news 184, 185
    Science/Nature News 192
    sources, stories and scoops 183
    see also Radio 4; World Service
Beckman Institute 175
Belton, Neil 173
Bernal, J. D. 156, 157
Berners-Lee, Tim 68, 71
bias in peer review 119, 121–2
Bibliography on
    Interdisciplinarity 66
bibliometrics 9
Big Idea, Ayrshire 204
Biological Abstracts 102
BioMed Central 69, 90
biotechnology, patents 43, 44–5
Birt, John 188
Blair, Tony 138, 142
blinded peer review 119
blogosphere 72
blogs 13–14, 71–2, 80, 91, 229
    controversies: stem cell research
        140
    culture scientifique 160
    physics 68, 72–4, 79
    and journalism 74–6
blooks 71, 152
Blumwald, Eduardo 37, 38
'Bologna Process' 63
books 161–2
    blog-inspired (blooks) 71, 152
    cross-disciplinary
        communication 60
    electronic 102–3, 152
    fiction *see* novels
    historical context 85, 89, 99
    importance 151–2, 157–61
    novels *see* novels
    'within' science 152–7
boundary work 6

bovine spongiform
    encephalopathy (BSE)
    135
Boyd, William 167
Boyle, Robert 26
Brahe, Tycho 67
Brewster, Sir David 198
British Council 215, 223
British Medical Journal (BMJ) 118,
    121, 127
Brockman, John 160
Bronowski, Jacob 157, 160
browsers 102
Brunner, John 169
BSE (bovine spongiform
    encephalopathy) 135
Budapest Open Access Initiative
    105, 108, 109
bureaucratic boundaries
    between disciplinary
    departments 60
Burma, internet 93
Bush, Vannevar 19–20, 23
business of scientific
    communication 11–13,
    67
    'small business' model 10–11

C

Cafés Scientifiques 209–10
    evolution 212–15
    formats 220–3
    future developments 224
    junior cafés 215–18, 219
    origins 210–12
    speakers, topics and audiences
        218–20
    web site 225
Canada, Cafés Scientifiques 223
Carson, Rachel 160, 161
Casti, John 172
cave paintings 84
CD-ROMs 101, 102
Centre for Culture, Science,
    Technology and Industry,
    Pole University of Lyon
    (CCSTI) 215, 216

CERN
  arXiv 68, 103–4
  ATLAS 77
  Large Hadron Collider (LHC)
    73–4, 75, 76
  open access publishing 68, 69
  socializing online 78
  World Wide Web 68
chaos theory 166, 168
Chemical Abstracts 99, 100, 152
Children's Science Book Review
  165
China
  internet 93
  nanotechnology 63
citations 7–8
  bibliometrics 9
  impact factors 8, 124
  indexes 100
  online journals 70
CiteULike 77
cloning 137
co-authorship 11
codex 84, 89
cognitive aspects of a discipline
  56
cognitive strategies for improving
    cross-disciplinary
    communication 57–9
collaboration
  online science 86
  see also cross-disciplinary
    communication
Comment on Reproduction Ethics
  141
Committee on Publication Ethics
  (COPE) 129
communism/communalism, as
    science norm 5, 25
  constitutive communication 27
community in science 3–4
  books 153, 154, 155
  business of scientific
    communication 12
  global 14–15
  origins 4–6
conferences 8
  informal communication 9
  international 11
  press 185–6
  proceedings 8, 153
Connotea 77
consensus *see* controversy and
    consensus
constitutive communication
  27–8

controversy and consensus 131,
  229
  from the particular to the
    general 131–3
  persuasion and public dialogue
    135–7
  public engagement 144–5
  radio programmes 185,
    186–7
  scientists' role in public
    controversies 133–4
  stem cell research 137–43
Conway, John 75
Cooper, William 173
Copernicus, Nicolas 67
Copernicus Publications 90
copyright issues, online science
  88
Cornell University, arXiv 12
corporate hospitality, *Our
    Dynamic Earth* 203
Cosmic Variance, 72, 73, 75
Council for Science and
    Technology (CST) 33
Cousteau, Jacques 160
Covington, Simon 'Syms' 172
CRC Handbooks 152
creationism 187
creativity index 70
creoles, development of 58
Crichton, Michael 168–9, 170,
  173
cross-disciplinarity 57
cross-disciplinary communication
  10, 53–4, 64–5
  cognitive strategies 57–9
  disciplines and their
    relationships 56–7
  growth and disciplinary
    fragmentation of science
    54–6
  internet 94
  nanotechnology 61–4
  online journals 70
  social strategies 59–61
culture scientifique 158, 159–60
cumulative research, and
    intellectual property 45
cuneiform 84
cybrids 137–43

D

Dallas, Duncan 210–11, 212, 213,
  216, 220, 224
Darwin, Charles 159, 171, 172

Darwinian evolution 187
data bases
  online 70, 100–1, 102
  patent searches 42
  peer review 124
Dawkins, Richard 134, 158, 161,
  187
Day After Tomorrow, The 168
deficit model of science
    communication 136
de Kruif, Paul 156, 158, 159
Deliberative Meeting of Citizens
  (Democs) 226
Denmark, Cafés Scientifiques
  221–2, 223
developing countries, open-access
    journals' publication fees
  107
DIALOG 100
dialogical communication 27,
  28–9
dialogic turn 228
digital divide 85, 93, 227
Digital Planet 190
Dijkstra, Edsger W. 117
*Directory of Open Access Journals
    (DOAJ)* 104, 111
disciplines
  cognitive aspects 56
  fragmentation of science
    54–6
  relationships between
    56–7
  social aspects 56–7
Discovery 179
disinterestedness, as science
    norm 5, 25
  constitutive communication
    27
disseminative communication
    27, 28–9, 30
Djerassi, Carl 168, 172, 173
Donaldson, Liam 142
Dorigo, Tomassio 74, 75
Dos Passos, John 169
drought resistant, GM plants
  37–42
DSpace 104
Dubos, Rene 156
Dudman, Clare 171

E

e-books 102–3, 152
Economist, The 75
Ecsite 208

Edelman, Gerald 175
edge.org 160
Edinburgh
    International Science Festival
        205
    *Our Dynamic Earth see Our
        Dynamic Earth*
education
    broadcasting 180
    Cafés Scientifiques 215, 217
    cross-disciplinary 59, 64
        nanotechnology 62–3, 64
    Our Dynamic Earth, 199–200,
        203, 204, 205
    textbooks 154–5
electronic networks 13, 101
Encyclopaedia Britannica 76
*Encylopaedia of Astronomy and
        Astrophysics* 102
English Civil War 5
English language 11, 93, 227
environmental issues *see Our
        Dynamic Earth*
e-periodicals *see* journals, online
e-prints 68
e-Prints software 105
e-Prints Soton 105
ethical codes 21–3, 25–6, 27,
        30, 31
    constitutive communication
        27–8
    disseminative and dialogical
        communication 29
    strategic communication 28
ethical issues, patents 46–7
ethical norms in science 26–9
EurekAlert! 188
European Association of
        Science Editors (EASE)
        119, 124
European Patent Office (EPO)
        42, 49
European Union
    'Bologna Process' 63
    nanotechnology 63
Evans, Lyn 74
evolution 187

F

Facebook 78
Farming World, The 179
Fermilab 74, 75
festschriften 153, 158, 227
Feynman, Richard 62
fiction *see* novels

films
    culture scientifique 160
    The Day After Tomorrow 168
    Jurassic Park 168
    The Skeptical Environmentalist
        168
Fitzroy, Robert 171, 172
Fleischmann, Martin 100
flickr 77
Flint, Caroline 142
Fox, Fiona 143
Fox Keller, Evelyn 161
fragmentation of science 54–6
France
    Cafés Philosophiques 210
    Cafés Scientifiques 211
        format 222
        junior cafés 215–16, 218
fraud, science 23
    internet 88
    peer review 120–1, 126
Frayn, Michael 173
Freud, Sigmund 159
Friends of the Earth 135

G

game theory 174
Garfield, Eugene 124
gender bias in peer review 122
genetically modified (GM) crops
    GM Nation? exercise 135–7
    patents 37–42, 46
George, Clare 167
Germany
    junior Café Scientifique 218
    Nazi era 24
    patents 43
Ginsparg, Paul 68
Glasgow Science Centre 204
Gleick, James 158, 162
global science community 14–15
Gödel, Kurt 171
Goetzmann, William 156
Golbeck, Jennifer 78
Golden Rice project 45
Google Scholar 70
    patents 42
    peer review 124
Gould, Stephen Jay 159
government web sites 91
Greenpeace 87
grey literature 153–4
growth of science 54–5
Guardian 148, 192
Gutenberg press 67, 79, 85

H

Haigneré, Claudie 216
Haldane, J. B. S. 172
Haldane principle of research
        council autonomy 23–4
Harris, Evan 141
Hart, Ernest 118
Hart, Michael 103
Hawking, Stephen 157
Hayles, Katherine 166
h-b index 70
Helling, Robert 77
Herrnstein, Richard 161, 162
Hess, Harry 153
Hewlett Packard (HP) 104
Higgs boson 74–5
*h-*index 70
Hirsch, Jorge 70
Hitchcock, Julian 143
Hogben, Lancelot 156, 157, 158,
        159
Holbrow, Charles 154
Hollywood films *see* films
Holyrood Brewery Foundation
        195–6
Human Fertilisation and
        Embryology Authority
        (HFEA) 137–8, 142, 143,
        148
Human Genetics Alert (HGA)
        140–1
human stem cell research *see*
        stem cell research
Hutton, James 198, 206
Huxley, Julian 156
hyperlinks 72

I

impact factors 8, 124
indexes 99–100, 102, 103
informal communication 9–10
Information Networking News
        104
In Our Time 190
INSPEC 102
Institute of Electrical and
        Electronics Engineers 63
Institute of Physics 69
    *Journal of Physics: Conference
        Series* 104
    Nanotechnology 63
institutional web sites 91
intellectual culture, books' role in
        155–6

intellectual development of
        science, books' role in 158
intellectual property (IP) 228
    nature of 35–7
    online science 88
    see also patents
interactivity, World Wide Web
        71–2
interdisciplinarity 57
interdisciplinary communication
        10, 53–4, 64–5
    cognitive strategies 57–9
    disciplines and their
        relationships 56–7
    growth and disciplinary
        fragmentation of science
        54–6
    internet 94
    nanotechnology 61–4
Intergovernmental Panel on
        Climate Change (IPCC) 91
International Committee of
        Medical Journal Editors
        (ICMJE) 130
international communication 11
International Federation of
        Library Associations (IFLA)
        105
International Science Shop
        Network 226
international web sites 91
internet 83–4, 95–6, 98, 108–9,
        228–9
    academic record 102–3
    advantages and disadvantages
        86–90
    and books 152–3
    business of scientific
        communication 12–13
    Cafés Scientifiques 222–3
    culture scientifique 160
    development 68, 98, 101–2
    epistemological issues 93–4
    fragmentation within online
        tools 15
    grey literature 154
    historical context 84–6, 97,
        100–2
    major forms of e-science 90–2
    open access 69, 94–5, 103–6
    patent searches 42
    peer review 125
    physics 67–8, 79
        blogging 72–6
        interactivity 71–2
        shaping knowledge 69–71

socializing 77–8
traditional print journals,
        challenging the 69
wikis 76–7
political/cultural dimensions
        92–3
praxis, issues of 94–5
radio broadcasting 188
researcher's perspective
        106–8
interviews, radio 183, 188–9
Iran
    Cafés Scientifiques 213
    internet 93
Ishiguro, Kazuo 170, 173

J

JANET 101
Japan
    Atom Technology Project 61
    Cafés Scientifiques 213, 214
        format 221
        junior cafés 218
    nanotechnology 61, 63n.3
    patents 44
jargon
    development for cross-
        disciplinary
        communication 58
    reduction in online journals
        70
Jensen, Pablo 211
Journal Citation Reports 124
journalism, science
    and blogging 74–6
    radio 178–90
Journal of Medical Internet Research
        (*JMIR*) 104, 108
Journal of Physics: Conference Series
        104
*Journal of the American Medical
        Association* (*JAMA*) 116,
        118, 122, 125
journals, science 6–7, 14, 15, 83
    and blogs 72, 79
    business of scientific
        communication 11–13
    citations 7–8, 124
    editorial process 7
    functions 7
    historical context 67, 99–100
    impact factors 8, 124
    information age 69, 79, 94–5
    interdisciplinary 63
    international 11

multidisciplinary 57, 59–60
    nanotechnology 63
    online 12–13, 69–71, 86, 90,
        95, 102–3
        advantages and
            disadvantages 87–8
        growth in East Asia 15
        open access 12–13, 69, 95,
            103–9
        peer review 125–6
        publication fees 107
    and patent applications 37,
        38–42
    pay journals 125
    peer review *see* peer review
    press releases 10
    scientists publishing research
        8, 10, 11
    types 7
JSTOR 103
Jung, Carl 159
junior Cafés Scientifiques 215–18,
        219
Jurassic Park 168

K

Kaku, Michio 78
Kehoe, Brendan 103
King, Sir David
    stem cell research 142
    'Universal ethical code for
        scientists' 21–3, 25–6, 27,
        30, 31
        constitutive communication
            27–8
        disseminative and dialogical
            communication 29
        strategic communication
            28
Krauss, Lawrence 72–3

L

LabLit 172–3, 175, 177
Lancet, The 116, 120, 127
language issues
    cross-disciplinary
        communication 58, 59
    English 11, 93, 227
    internet 93
    Mandarin 15
    technical 55
Large Hadron Collider (LHC)
        73–4, 75, 76
Laurence, William 186

Lauterbur, Paul 117
'leaked' information 88
learned societies 5
    journals 6–7
        nanotechnology 63
        peer review 118, 123
Leavis, F. R. 155–6
Leavitt, David 167
lecture-demonstration 4
Lessing, Doris 166
Lessing, Lawrence 160
Levi, Primo 159
Levin, Janna 171–2
Lewis, Sinclair 156
liability for patent infringement
    45
librarians 99, 101, 102
Lidar, Daniel 72
life science patents 43, 45
Lightman, Alan 172
Lisi, Garrett 73
literacy, scientific 94
Literary and Philosophical
    Society of Manchester
    118–19
Living Knowledge 226
Lomborg, Bjorn 169
Loring, J. 46
Los Alamos Physics Archive 103

M

McClintock, Barbara 161
McCormmach, Russell 170, 173
Macdonald, Roger 172
McEwan, Ian 173–4, 175
Mandarin 15
masked peer review 119
Massachusetts Institute of
        Technology (MIT) 104
materials science 55
Material World 190
MATICO 100
media, scientists' communication
        with 10, 139, 144
    stem cell research 142
    see also journalism, science;
        specific forms of media
mediation strategy, cross-
        disciplinary
        communication 59
    nanotechnology 64
Medical Journal of Australia (MJA)
        125
Medical Research Council 24,
        138

Merck Manual 153
Merton's 'Normative structure of
        science' 5, 23–6, 27, 30
    constitutive communication
        27, 28
    dialogical communication
        28–9
    disseminating communication
        29, 30
    intellectual property 44
    strategic communication 28
Millennium Commission 199,
        203, 204, 205
Miller, Hugh 198
Minger, Stephen 137, 139
Mitchell, David 170, 173
Mixed States 72, 80
modularization strategy,
        cross-disciplinary
        communication 58
Mooney, Chris 160
Moore, Patrick 179
Mosaic 102
*Mrs. Marcet's Conversations on
        Chemistry* 154
Mullis, Kary 168
multidisciplinarity 57
    internet 94
Murray, Charles 161, 162
MySpace 78

N

Nakamura, Masaki 214
Naked Scientists 165, 192
Nanoforum 66
Nano Letters 63
Nanotechnology 63
nanotechnology
    cross-disciplinary
        communication 56,
        61–4
    public engagement 30–1
    upstream engagement 30
    web sites 33, 66
Nanotechnology Now 66
NASA 215
NASA RECON 100
Nature 7
    'Climate Feedback' site 86–7
    drought resistant GM plants
        37, 38
    internet 69
        forum 97
    multidisciplinarity 57, 59
    peer review 117, 125, 130

    press releases, radio coverage of
        183, 186
    subject indexes 99
    wikis 76
Nature Biotechnology 42
Nature Notebook 179
Nazi German 24
Nemo, Amsterdam 202
Netscape 102
network technologies 13, 101
New England Journal of Medicine
        120
New Journal of Physics (NJP) 69
New Metropolis, Amsterdam 202
New Scientist 74, 75, 79
news coverage of science
    blogs 72, 73, 74–5
    controversies
        obesity, 'health benefits' of
            132–3
        stem cell research 138–9,
            140
    drought resistant GM plants
        38, 42
    journals 10
    radio 183–6
New Yorker 159
New York Times 186
norms of science 5, 23–6, 27, 30
    constitutive communication
        27, 28
    dialogical communication
        28–9
    disseminating communication
        29, 30
    intellectual property 44
    strategic communication 28
North Carolina science blogging
        conference 80
Not Even Wrong 73–4, 76
Nova ScienceNow 215, 219, 220
novels 157
    facts in 168–70
    LabLit 172–3
    McEwan, Ian 173–4
    Powers, Richard 174–6
    reading science 166–7
    science in context 170–2
    use 167
Nuclear Science Abstracts 100

O

obesity, 'health benefits' of 131–3
online journals *see* journals,
        online

online science *see* internet
open-access publishing 12–13, 68,
      69, 94–5, 103–9
*Open*DOAR 111
open peer review 71, 125
Optics Express 69
ORBIT 100
organized skepticism, as science
      norm 5, 25
Ostwald, Wilhelm 60n.2
Our Dynamic Earth, Edinburgh 207
   background 195–6
   commercial imperative 203
   development 199–203
   funding 198–9, 203–4, 205
   refreshing 205–7
   roots 196–8
   Scottish Science Centre
         Network 204–5
   web site 208
Oxford University, *Oxford Research
      Archive* 105
Oxford University Press 105
Ozimic, Anthony 141

P

Palestine, Café Scientifique 223
Paris Review 177
patents 35, 47, 228
   case study: drought resistant
         GM plants 37–8
      paper and patent application
            compared 38–42
   commercialization of research
         20
   ethical dimensions 46–7
   intellectual property, nature of
         35–7
   searches 42–3
   universities 43–6
Patents Act 1977 (UK) 37
pay journals 125
peer review 7, 15, 99, 115, 127,
      228
   business of scientific
         communication 12
   costs 12
   effectiveness issues 120–1
   fairness issues 121–2
   future 126–7
   historical context 118–19
   nature of 115–17
   new developments in 124–6
   norms of science 5
   online 88, 104

open 71, 125
   process 119
   reasons for 122–4
personal web sites 92
*Philosophical Transactions of the
      Royal Society* 67, 99, 118,
      153
Physical Review Letters 70
physics 67–8
   blogging 72–4
      head to head with journalism
            74–5
      as reporting 76
   interactive web 71–2
   reductionist strategy, cross-
         disciplinary
         communication 58
   shaping knowledge online
         69–71
   socializing online 77–8
   surviving the information age
         79
   textbooks 154–5
   traditional model, challenging
         the 69
   wikifying the web 76–7
Physics World 74, 76, 77, 79
PhysMath Central 69
pirating
   books 153
   online 89
plagiarism
   internet 88
   self-plagiarism 15
plate tectonics 153
plays 173
PloS One 126
podcasts 92, 178, 189, 190
polymerase chain reaction (PCR)
      168
Powers, Richard 174–6
preprints 68, 90
   and books 152–3
   see also arXiv
press conferences 185–6
press releases
   and patent applications 37–8
   radio coverage 183–4, 186,
         188
Priestley, Joseph 60n.2
printing press 67, 79, 85
print media 98, 99–100
   see also books; journals, science
*Proceedings of the National
      Academy of Sciences* (*PNAS*)
      38–42

professional web sites 90–1
'professor's privilege' 43
Project Gutenberg 103
publication of research 8–11
   intellectual property 44
   see also books; journals, science;
         publications
publications
   gatekeeping 6, 7
   impact 67
   origins of community in
         science 5–6
   see also books; journals, science
publishing teams 11
public culture, books' role in
      156–7
Public Library of Science 90, 126
PubMed 42, 124
Pulitzer Prize 156

Q

Quantum Diaries 73
Quantum Pontiff 72
Quintaville, Josephine 141

R

radio and other audio 178
   controversy 186–7
   culture scientifique 159
   history 178–80
   information age 188–90
   listeners 181–3, 189–90
   nature of 180–1
   news 184
   sources, stories and scoops
         183–4
   texture and colour in sound
         189
   topicality 185–6
Radio 4: 182, 190, 192
Ramanujan, Srinivasa 167
Rankin, Sir Alan 195
recruiting people into science,
      books' role 158, 159
ReDiscover Fund 205, 206
reductionist strategy, cross-
      disciplinary
      communication 58
Registry of Open-Access
      Repositories (ROAR) 111
Rennie, Drummond 122–3
repositories, research 104–5, 107,
      109
   registry (ROAR) 111

reputation issues
  disciplinary fragmentation of
    science 55
  online science 89
  patent applications and articles
    compared 41
  peer review 116, 123
Research Assessment Exercise (UK)
  archives 105
  bibliometrics 9
  peer-reviewed publications 99
research exemption in patent
  regimes 44–5
research institutions
  cross-disciplinary
    communication/interdisci
    plinary research 60, 64
  nanotechnology 63n.3
  patents 43
Rethinking Interdisciplinarity 66
reward structure in science 28
ROAR (Registry of Open-Access
  Repositories) 111
Roberts, Adam 168
Roco, Mihail 62
Rorvik, David 169, 170
Royal Society
  in fiction 170
  internet 87
    open access 105
  peer review 117, 118, 119
  *Philosophical Transactions of the
    Royal Society* 67, 99, 118,
    153
  stem cell research 138

S

Sacks, Oliver 159, 175, 212
Sagan, Carl 156, 157, 159, 160
Samuelson, Paul 155
Satrosphere, Aberdeen 204
Sautet, Marc 210
Schön, Jan Hendrik 120
Schrödinger, Erwin 172, 173
Scibooks 165
SciDev.Net 188
Science 7
  internet 69
  multidisciplinarity 57, 59
  peer review 120
  press releases, radio coverage of
    183, 186
  web site 165
Science and Industry 179
scienceblogs.com 160

science centres 202–3, 204, 205
  see also Our Dynamic Earth
Science Citation Index 100, 152
Science Commons 18
Science in Action 179, 190
Science Media Centre (SMC) 144,
  148
  stem cell research 139, 142,
    143
scientometrics 9
SCIRUS
  patents 42
  usage 107, 108
SciVee 78
Scopus 70
Scottish Science Advisory
  Committee 204, 206
Scottish Science Centre Network
  204–5
scrolls 84
Seed Magazine 160
self-archiving 12–13
self-plagiarism 15
Sensation, Dundee 204
SETPOINTs 208
Shakespeare, Tom 213, 216
SHERPA project 105, 111
SigmaXi 215
Silver, Lee 169–70
Silverplatter 102
simplification strategy, cross-
  disciplinary
  communication 58
Skeptical Environmentalist, The
  168
'small business' model 10–11
Smith, Zadie 170, 173
Snow, C. P. 155–6, 159, 160,
  172
Sobel, Dava 159, 170
social aspects of a discipline
  56–7
social contract for science 19–20,
  30–1
  ethical norms in science 26–9
  King's 'Universal ethical code
    for scientists' 21–3
  Merton's 'Normative structure
    of science' 23–5
socializing online 77–8
social networking sites 78
social relations of science
  movement 24
social strategies for improving
  cross-disciplinary
  communication 59–61

social systems 4
  see also community in science
social tagging 77
Society of Unborn Children 141
sociology of scientific knowledge
  6
Solerby, José 70
Southampton University, *e-Prints
  Soton* 105
South Korea, nanotechnology 63
Stanford Linear Accelerator
  Center (SLAC) 68
Stanley Robinson, Kim 173
stem cell research
  controversy and consensus
    137–43
  patents 46
STEMNET 208
Stoppard, Tom 173
strategic communication 27,
  28
string theory 13, 73–4
Sudbø, John 120
Summit, Roger 100
Swedish Medical Research Council
  122

T

textbooks *see* books
Transactions on Nanotechnology
  63

U

United Kingdom
  GM Nation? exercise 135–7
  Research Assessment Exercise
    archives 105
    bibliometrics 9
    peer-reviewed publications
      99
'Universal ethical code for
    scientists' 21–3, 25–6, 27,
    30, 31
  constitutive communication
    27–8
  disseminative and dialogical
    communication 29
  strategic communication 28

V

Vesalius 85
video-conferencing, Cafés
  Scientifiques 223

W

Wallace, Jim 204
Waterston, J. J. 117
Watson, James 60n.2, 133–4
    books 155, 158, 159, 161
webcasts, Cafés Scientifiques
    223
web logs *see* blogs
web sites *see* internet
WebSpirs 102
Wegener, Alfred 171
Weizenbaum, Joseph 158
Wellcome Trust 148, 205, 213,
    216, 224
WGBH channel 215, 221
Whitehouse, David 186

Wikipedia 13, 76, 87
    physics 68, 77
    review methods 106
Wilkinson, James 185
Wilmut, Ian 139
Wilson, E. O. 155, 158
Winterson, Jeanette 167
Wired 75
Wittgenstein, Ludwig 172
Woit, Peter 73–4, 76
Wolfson Foundation 205
Woo-suk, Hwang 23, 120
World Intellectual Property
    Organization (WIPO) 49
World Nuclear Association 87
World Service
    information age 188, 190

listeners 181–3
    news 184
World Summit on the Information
    Society (WSIS) 97
World Trade Organization, TRIPS
    agreement 37
world wide web *see* internet

Y

YouTube 78, 92
'yuk factor' 140

Z

Zanardi, Paolo 72
Zimmer, Carl 160, 170n.3